COLLABORATIVE RESEARCH IN THEORY AND PRACTICE

The Poetics of Letting Go

Kate Pahl, Richard Steadman-Jones
and Lalitha Vasudevan

*with Hugh Escott, Cristina Salazar Gallardo, Kristine
Rodriguez Kerr, Andrew McMillan, Jonathan May, Katie
Scott Newhouse, Steve Pool and Vicky Ward*

I0223064

BRISTOL
UNIVERSITY
PRESS

First published in Great Britain in 2023 by

Bristol University Press
University of Bristol
1-9 Old Park Hill
Bristol
BS2 8BB
UK
t: +44 (0)117 374 6645
e: bup-info@bristol.ac.uk

Details of international sales and distribution partners are available at bristoluniversitypress.co.uk

© Bristol University Press 2023

British Library Cataloguing in Publication Data
A catalogue record for this book is available from the British Library

ISBN 978-1-5292-1509-0 hardcover
ISBN 978-1-5292-1510-6 paperback
ISBN 978-1-5292-1511-3 ePub
ISBN 978-1-5292-1512-0 ePdf

Cover design: Nicky Borowiec
Front cover image: Adobe Stock / Aleksander Savelev

Contents

Notes on Collaborators

Hugh Escott is Senior Lecturer in English Language at Sheffield Hallam University. His research focuses on how language and literacy are understood in everyday contexts, and how individuals negotiate or subvert institutional conceptualizations of language and literacy. This often involves taking an interdisciplinary approach to research, and working with children and young people to explore how they play with language and literacy in classrooms, creative writing workshops and informal contexts.

Jonathan May is a British artist, writer and cultural producer living between London and Yorkshire. His work spans photography and live practice. His photography is a reflective process, used to construct moments of stillness and introspection and often paired with bookmaking to create small runs of intimate photographic artefacts. He has an MA in Performance from Goldsmiths University of London and is a Clore Fellow. He has previously led cultural organizations and creative programs with Abandon Normal Devices, British Council and the London International Festival of Theatre. He is Executive Producer of the experiential art collective Marshmallow Laser Feast.

Andrew McMillan's first collection, *physical*, was the first poetry collection to win the Guardian First Book Award; it also won a Somerset Maugham Award, an Eric Gregory Award, a Northern Writers' Award and the Aldeburgh First Collection Prize. In 2019 it was voted as one of the top 25 poetry books of the past 25 years by the Booksellers Association. His second collection, *playtime*, won the inaugural Polari Prize. He is a senior lecturer at the Manchester Writing School at Manchester Metropolitan University and is a fellow of the Royal Society of Literature. *pandemonium*, his third collection, is out now.

Kate Pahl is Professor of Arts and Literacy at Manchester Metropolitan University. Her work is concerned with literacy in communities. Her projects have included research on zine-making, trees and the natural world, and work with youth using arts methods. She is also the author, with Jennifer Rowsell, of *Living Literacies: Literacy for Social Change* (2020).

Steve Pool is a visual artist who works with people to help realize ideas. He originally trained as a sculptor. He now works across different media, including video, sound and projection mapping, and writing. On most of his projects he usually ends up making something. He has been artist-in-residence on a number of research projects and is currently working towards his PhD exploring the concept of the artist's residency as method.

Kristine Rodriguez Kerr is Academic Director and Clinical Associate Professor for the MS in Professional Writing program at New York University School of Professional Studies. Her research interests include new literacy studies, educational justice, multimodal composing, participatory culture, and the teaching of writing in both traditional and non-traditional spaces of learning. She earned an EdD from Teachers College, Columbia University and an MA and BA from Carnegie Mellon University.

Cristina Salazar Gallardo is a qualitative researcher based in Mexico. She is interested in the intersection of ethnographic enquiry and digital practices. She currently teaches at the Center for Research and Teaching in Economics and the Virtual University of Guanajuato.

Katie Scott Newhouse is a Bruce S. Goldberg Postdoctoral Fellow in Youth Wellbeing at Teachers College, Columbia University in the Media and Social Change Lab. Her fields of study include Disability Studies in Education, Disability Justice and Critical Spatial Theory. Her research focuses on the experiences of youth mandated to attend specialized restrictive programs (such as special education and juvenile justice) and the spatial implications of enrolment in these programs.

Richard Steadman-Jones is Senior University Teacher in the School of English at the University of Sheffield. He has developed a range of research on the linguistic dimensions of intercultural encounters and is also interested in narrative, digital media, visual art and myth.

Lalitha Vasudevan is Professor of Technology and Education, Vice Dean for Digital Innovation and Managing Director of the Teachers College Digital Futures Institute, and Director of the Media and Social Change Lab. Her work examines communication and new media, adolescents' literacies, youth media and multimodal storytelling in community-based settings.

Vicky Ward has extensive experience in working collaboratively within communities. As a qualified social worker and teacher, her work is rooted in good practice across community development, community education, and social work methods and approaches. Vicky has also worked with socially engaged arts practices through photography and is interested in how art can increase voice and shape ways of seeing while helping us think critically about issues of power and whose knowledge counts when working within communities.

Acknowledgements

Richard and Kate would like to thank the 'Language as Talisman' team, especially Robin Bone, Deborah Bullivant, Hugh Escott, Jane Hodson and David Hyatt; the 'Communicating Wisdom' team, especially Marcus Hurcombe and Johan Siebers. Also, our funders, the Arts and Humanities Research Council (AHRC) and the Economic and Social Research Council.

Kate also would like to thank her colleagues, especially Khawla Badwan, Abi Hackett, Mel Hall, Rachel Holmes, Keri Facer, Gabrielle Ivinson, Deborah James, Peter Kraftl, Maggie MacLure, Caitlin Nunn, Frances Press, Zanib Rasool, Harriet Rowley, Jennifer Rowsell and Sarah Truman and the project teams in the 'Odd: Feeling Different in the World of Education' team (AHRC), the 'Questioning the Form' team (Arts and Humanities Research Council/Global Challenges Research Fund) especially Gloria Kiconco, and the 'Voices of the Future' team (Natural Environment Research Council) and special thanks to the AHRC Connected Communities program and all the young people involved in her projects.

Richard would like to thank his students.

Lalitha, Kate and Richard would like to thank Catherine Cheng-Stahl, for being the first reader of our whole manuscript and for her care-filled close reading and keen editorial eye. She has made this book better with her generous comments and suggestions.

Lalitha would like to thank the young people who have been engaged interlocutors and allowed her to learn and inquire with them; the adults in their lives who welcomed her and her research teams into their institutional spaces; her students, who continue to inspire, energize and play in gloriously unexpected ways; and her colleagues who made time and space for wanderings and wonderings, especially Lesley Bartlett, Katie Conway, David Hansen, Nathan Holbert, Sonali Rajan, Anne Burns Thomas, Beth Tipton, Erica Walker, Haeny Yoon; the 'Reimagining Futures' team, especially Sharieff Clayton, Ana Dopazo, Eric Fernandez, Melanie Hibbert, Kristine Rodriguez Kerr, Katie Newhouse, Ahram Park and Cristina Salazar Gallardo; and the Media and Social Change Lab, especially Ioana Literat, Detra Price-Dennis, Joe Riina-Ferrie, and Azsaneé Truss.

Prologue

As we begin this book, we thought it might be helpful if we prepared our audience. This is not a 'methods' book, but it does discuss methods. It is not a 'how to' book but it talks about mechanisms and ways of doing things. It is not trying to produce new paradigms but it does engage with unusual ranges of literature, readings and other texts. The book is about the underbelly of research. It describes the thinking that underlies certain kinds of research or the 'headnotes' that research generates (Ottenberg, 1990). We highlight an approach to research that we describe as the 'poetics' of research, which is concerned with how we work and the artfulness of that work in a research project.

Our determination to write this book is connected to our sense that there was something different about the projects we describe in the pages that follow, and that this difference was also opaque and hard to articulate. We have long been concerned that the rules or the procedures of the academic space can limit what is possible within co-produced, participatory research projects (Bell and Pahl, 2018). The question, in a sense, is one of attention. The formal outcomes of such research are not always what is most interesting about it and hence we suggest moving attention to something a bit different. We believe that some of the 'mess around the edges' of this work has not been captured adequately, or alternatively not seen as the 'work' of the projects. Creating living knowledge (Facer and Enright, 2016) with communities requires an understanding of where the knowledge is situated, and where the research attention lies. Our aim in this book is to produce a refocusing of attention on some aspects of research that might get missed. This could include an attention to poetic, embodied and enchanted elements of the work. We have written chapters about approaches such as enchantment and hypertext as a way of signalling a more arts and humanities approach to the journey of doing research, not in contrast to, but alongside and in conversation with, social science practices.

The form of the book is also a bit unusual. We used the term 'poetics' as a way of signalling what is distinct about the work explored here. It does not seem appropriate or adequate to discuss the work in the conventional language of methodology or methods. To recognize the importance of the stuff at the edges one needs a kind of judgement that does not anticipate anything in advance but is responsive to the moment. This book experiments with form in that writers approach their topic both using academic prose and also using more innovative and poetic modes. This signals the shift we want to encourage, from a formal academic social science model of discovery to a model that allows for opacity and indeterminacy as well as some other elements such as worthiness. Reading the book may therefore be slightly unsettling as the voices within it move and change. Included in the list of chapters are 'Interludes' that reflect on research from different perspectives and voices, providing a counterpart to our authorial contributions. Additionally, this book explores how research practices can sometimes produce different kinds of writing that can happen in ways that may appear esoteric or veiled. For instance, community–university partnerships need to be described in language that crosses both spaces. This might mean using different conceptual tools that come from community contexts, and not merely dropping them into literature reviews or conceptual frameworks, but letting go of both to consider what new directions and ways of being emerge in response. One implication of this proposed shift is for universities to learn to be a bit more humble with their role in research and, in turn, researchers could then attune to the spaces where research is being carried out. Thus, the work described in this book attempts to make sense of this conundrum from multiple angles.

While crafting this book, we conversed with collaborators and mentors and threads of these conversations and remembrances have found and wound their way into the pages that follow. The term 'poetics' is used to signal both the style of writing – which is literary and evocative in some places, but also an approach that recognizes the artful and slightly different tempo of the work and the thinking, at the time these projects took place and in the course of our collective writing. As such, this book has a rhythm that moves with the research process, and acknowledges the different voices within it. The word 'poetics' also allows us to recognize and call attention to the ways that research can be communicated in different registers and that different relationships can exist between these different registers. In this way, the book celebrates reading as an act of discovery. For example, the 'Hypertext' chapter explores how the juxtaposition of different kinds of texts can create new resonances for the researcher. This is characteristic of the book as a whole as different voices jostle in the chapters and interludes, producing a model for polyphonic writing and reading.

We hope that by opening up the space of research to other forces and discourses, research itself will change. The following quotation from Ralph Waldo Emerson urges us to witness and receive the passing of time and space, and to value small moments in the research process:

Dream delivers us to dream, and there is no end to illusion. Life is a train of moods like a string of beads, and as we pass through them they prove to be many-colored lenses which paint the world their own hue, and each shows only what lies in its focus. (Emerson, 2003)

We have strung together proverbial beads of reflections, insights and learnings from our collaborative endeavours. This book is an act of both looking back and gazing ahead at the horizon in front of us. We hope that as you forge new paths, raise new questions through and about research, and form new bonds with the communities to whom you are tethered through your work and words, that you are moved to add another bead or two to the string we have started here.

1

Introduction

Kate Pahl, Richard Steadman-Jones and Lalitha Vasudevan

Let us think about three spaces.

★★★

The first is a room on the second floor. To enter the room you need an identity card with a magnetic strip. Swipe it through the card reader and the doors to the building will open. But if you don't have an identity card, you must use the intercom and hope that an invisible someone will let you in. Some visitors are unsure what to say when the disembodied voice greets them from the entry system's speaker. "I'm here to see …". Sometimes they forget the name and have to rummage around in a bag for a letter or a note on the back of an envelope. The voice is civil but not especially welcoming. The onus is on you to explain why you are here.

Passing through the doors you enter an open space two storeys high. There are some chairs and tables to your right and a staircase directly in front of you. There is no indication of how you find the room that they mentioned. They said something about going up, so you begin to climb the stairs and halfway up you almost collide with the woman who has organized the event. "You managed to get in, then?" she says brightly, "I was worried they might have left you standing outside." She turns and heads back up the staircase, so you follow and speak to her back: "I wasn't sure if I'd got the right building."

★★★

Now we are at the edge of a pond. We are not in the depths of the picturesque countryside. The pond is a relic of a long industrial history and this is a location on the margins of the urban and the rural. If you turn away, you can see a housing estate and some big modern sheds that probably house a business of some kind. But across the pond on the other side are open fields, and to the right is a stand of willow trees. Their longest branches reach down and touch the surface of the water. It is engrossing to watch them shiver and tremble in the breeze. Beyond the fields there is a road, but you can't hear the traffic – or only very distantly.

A man sits by the pond on a folding stool. He must be around 60 but it's hard to tell. He has a big military-green umbrella and is surrounded by kit: a couple of plastic boxes with handles, a thermos flask, something wrapped up in foil, a fishing rod. On either side of him are two young people – a boy and a girl, 14, perhaps. All three are looking at his hands. For a moment his 60-something hands are the centre of the world. One holds a hook and the other a line. He passes the line through the eye of the hook, makes a loop, and then slips the loop over the hook and pulls it tight. He is ready to begin.

★★★

Finally we are on a city street. The sun is overhead. You could fry eggs on the paving slabs. Why are we standing on this corner? Because it's at the centre of a cluster of important places. If you turn left at the lights and walk a couple of blocks you come to the playground. Home is straight ahead. The city centre is behind us. And so the corner has itself become an important place, although not everyone knows this. Not everyone sees it for what it is.

Standing here, we watch the people passing. When someone we know comes by, we greet them in the way you greet a friend. We look down the long straight street towards the city. Today we are heading that way ourselves but we're waiting while a couple of people buy cans to drink on the journey, and we keep to the shade because the light is so bright and the sun is so hot. We talk while we wait – about the people passing, about where we're going. Someone makes a joke about the heat and we laugh. We laugh a lot together here.

★★★

And so now you arrive in the room where they are holding the event and there are eight or nine people there, no one you know. There are rows of chairs facing the front, a screen, a data projector. The woman who invited you introduces you to a young man with dark hair and glasses and you talk together in a stilted way while she disappears again and returns with a bottle of water. She announces that we're going to begin and everyone sits. As she speaks you look around you. The room is like a little cinema and they've even pulled down the blinds so it's dark enough to see the projected slides. You watch as one replaces another. You can't quite shake the feeling that you're at the movies. The expectation seems to be that you'll just sit here in the almost dark and listen.

The talk does not last as long as a film. When it comes to an end, the group agrees that they will divide up into threes and fours to talk. They move the chairs and maybe you're wrong but they seem a little excited as they break up the rows and regroup them into clusters, as if rearranging the furniture is some minor act of rebellion. They are interested in what you have to say. They listen respectfully. The larger group reforms and you say your piece, drink a cup of coffee, and leave. Getting out is easier than getting in. You just press the green button to the left of the door and it opens automatically to discharge you into the street outside, where two dogs are barking at one another, their owners rolling their eyes at the misplaced seriousness of the animals' confrontation.

★★★

Under the water is another world entirely. There are carp and tench and roach and bream. The fisherman understands the behaviour of each species – how they inhabit the space beneath the surface – but it's impossible to learn this through direct observation. There are books you can read. 'Tench are not lovers of strong light', writes Fred J. Taylor in *The Shell Book of Angling*, 'and it is reasonable to expect a big bait to be taken boldly during the early morning hours before they retire to the shade of the weed-beds, lily-pads or rush-beds as the sun grows higher' (Taylor, 1979, pp 159–160). But the boy and the girl are learning this stuff in situ from the man with the military-green umbrella. Together they imagine the world beneath the water – a world that is connected to ours only through the vector of the fishing line.

Their lives are not idyllic and neither, in truth, is the pond, although it's quiet and, to the fishers, compelling. It is a place of waiting.[1] They wait for a sign that the fish are feeding. For a tug on the line. And as a person waits, time thickens. It becomes viscous – like water in comparison to air. Here, time moves slowly and the willows shiver and the tench retire and the sun grows higher. For a while, the world is very still. The girl closes her eyes and breathes. The boy lets his imagination drop down the line and roam around among the weeds at the bottom. Who knows what the fisherman is thinking as he opens the foil package and takes out a sandwich. The line is suddenly taut. He begins to reel it in.

★★★

We have travelled into the city and here the streets are heaving with people. We can't all stay together, and our paths split and regather to form a braided pattern upon the sidewalks and the kerbstones. Our mood is high and rising. We shout to one another so the people in between us are caught in the middle of laughter. Some of them turn. Most hurry on. We take out our cameras. For a moment, I am at Fashion Week: "Give me fierce." She does. It's a good photograph. Our paths meet. I show her. She laughs.

---

[1]   'Waiting is an enchantment', says Roland Barthes in *A Lover's Discourse* (1978, p 38).

At the art museum we take more photographs. People flow through the rooms and the current carries us – to get your shot you have to wait until you're just in front of the picture. Our own stories spark against the stories of the photographs so that the air is thick with memories and, once or twice, I catch my breath as the detail of a picture conjures up the recollection of something that happened to me in the same place, on a similar street. Some of the photographs seem intended for me. Others for someone else. We drift along on the river of people and, in the cameras, we collect the stories that are ours and ours alone. We are surprised that no one challenges us and later we decide that they must have thought we were journalists writing a feature on the latest exhibition.

Falling into a reckoning

The writing of this book has been an excavation of sorts into data and stories from our collaborative projects. It has provided each of us with ample opportunities to dig into our memories and remember the shifts in thinking we experienced as we have travelled – in some cases several years – past the moments of collaboration we write about here. The descriptions in the previous section exemplify the kinds of memories we carry from each of our different projects.

Prior to coming together to compose this text, we each had worked in different locations around the world, with different communities, at different times in the last few decades, and often in very different contexts. Yet our shared observations about the nature of collaboration, community and partnerships hold our collective insights together within these pages.

We have read and written this book, inhabited its ideas, and conferred about its purpose and meanings. We have lived inside of our memories of the projects that serve as the foundation of the ideas that we share in these pages. We have exchanged and recommended texts to one another, read to each other excerpts of essays and poetry, and worked through bouts of writer's block that were caused not by an absence of ideas but by an abundance of immediacy that we felt about what needed to be said and shared before another idea or story could be layered on top of it. The stories tumbled out, crashing into one another, before merging into a shared narrative like discursive tributaries creating a river of reconfigured words and meanings.

One of the original working titles for this book was 'Untethered', a word that signifies the simultaneity of relative freedom away from weighted forces – like the university and schools whose relationships with communities are imbued with complicated histories – and recognition of their persistent presence. A glimpse of this notion remains in the book's subtitle with the phrase 'poetics of letting go', because our conversations about what we are trying to effect through our work and with this book continued to evolve throughout the writing of this text. The word 'poetics' provides an opening to see research as, in some ways, a poetic enterprise, with its own architecture that could include enchantment, failure and unknowing as much as a 'scientific' mode of enquiry. We relied on poetic forms of communication and drew on multiple genres and modes of expression in order to elucidate sensorial aspects of research like affect and memory that are not easily categorized. To quote from A.R. Ammons (1986), we:

> looked for the forms
> things want to come as
> ...
> not so much looking for the shape
> as being available
> to any shape that may be
> summoning itself
> through me
> from the self not mine but ours.

These decisions echoed those we made in the research projects we discuss throughout this book, each of us very much attending to the work of 'being available to any shape that may be summoning itself' through us.

In this chapter, we introduce some of the reasons that drove us to compose this book in the first place. The book is written to challenge a singular view of the university and to move towards more collaborative modes of enquiry. This involves working alongside the members of the communities and institutions where we locate our work. We recognize that such a shift is neither an easy one, nor one that is necessarily desired or plausible. Thus, we write from a hopeful stance. Our thinking pushes towards the 'not yet' of the hopeful future, drawing on ideas from philosopher Ernst Bloch (1986), particularly that of the 'concrete utopia'. The writing of this book is situated in a kind of 'forward dreaming' of what 'could be', as well as what is and what was.

The ideas we explore in this book emerged from a range of projects that took place between 2005 and 2020 (see Appendix). The projects were situated across different spaces and places and some of these were located within universities. Indeed, most of the work was funded by research grants that are only available to applicants in academic institutions. However, other spaces were equally, if not more, important when it came to the work itself. Much of the activity arising from the 'Communicating Wisdom' project, for example, took place on the edges of fishing ponds in the town of Rotherham, South Yorkshire (one of the spaces discussed in the opening vignettes to this chapter) and, in the same way, much of the work of the 'Education In-Between' and 'Reimagining Futures' projects took place in the city streets and office buildings and in public spaces such as art museums and courthouses (see vignettes). It is not that these spaces simply became the objects of enquiry for the employees of universities, a dynamic that is sometimes characterized in terms of 'extractive' research; on the contrary, the projects enabled work that was firmly located in those spaces and took its meaning from those contexts.

All the projects intentionally sought to unmake and remake traditional flows of power and authority in academic research – through method, practice, framing and so on – one result of which was that all parties experienced a form of vulnerability that comes with becoming visible in new ways to new people. Put another way, the projects rendered all parties involved – including community members, educators, youth, researchers, artists and other collaborators – vulnerable to being seen and known outside of the boundaries of the traditional roles of researchers and participants. Often, university-initiated research seeks to maintain flows of power that place the researcher in the seat of knowledge maker and participants as the people being researched, or as the 'beneficiaries' of research. In our work, and with this book, we seek to destabilize these seemingly sedimented, unidirectional flows of power.

The projects that we worked on had these qualities or characteristics:

- They were collaborative, involving interactions between academics, community groups, youth workers, children and teachers.
- They all involved artists in some way, but in some projects, artists were less present.
- They were concerned mostly, but not entirely, with young people and their meaning making practices.
- They drew on the humanities in terms of disciplinary orientation but used some methods from social science (for example, ethnography).
- They drew on visual and embodied ways of knowing and expressing, but some projects were mainly concerned with literacy and language.
- The projects were co-led, often with community researchers, but they did tend to be situated, in terms of funding and structures, in universities.
- The projects were concerned with generating new meanings, not disseminating existing ones, but in some cases, an element of historical validation was important.

The sites where the projects took place are evoked in the opening descriptions. Community members might feel uneasy in the university, and sometimes academics felt unsure when sitting and fishing by a pond. The position of the expert was looser, more porous, as projects surfaced knowledge in different ways. The project team might have been linked through common purposes, such as young people's well-being, but these purposes and intentions did not conform necessarily to traditional research questions. In these instances, rather than academics coming to the sites with ready-made research questions, the sites and spaces of the projects produced generative ideas that framed the projects from their inception, allowing questions to breathe and evolve as relationships between people, spaces and institutions did, as well.

The ingredients of the process often lay outside what are commonly understood as academic disciplines such as linguistics, history, anthropology and geography. Meanings were culturally determined by place, experience, history, and these are, of course, the subject as well as the object of disciplines. But in the moments of encountering and co-creating meaning, different things happen. People might articulate their meanings from complexities of experience not described or defined by the disciplines created to understand them. This is particularly pertinent in contexts that lie outside the experiential frame of many academics. Instead of the 'cultural stuff' of life produced within the domains of the experiential landscape of the academic, the meanings created might be more nuanced. They might cross languages, be transnational, local and global; they might be hybrid, co-created within conditions that are unfamiliar to those who attempt to describe them. This is why co-production is so important – it produces meanings from many voices and holds together many lived realities.

Three concepts anchor our discussions about these endeavours and how we have thought about the nature of collaboration. We begin with a discussion of research 'messiness', which was not incidental but rather a requisite feature of collaborative research. A reflection on 'temporality' follows, calling attention to the relevance of time and timeliness for the contexts of our research. 'Spatiality', which is entwined with discussions of time, rounds out our framing of an ethos for research collaborations with communities and institutions.

The messiness of research

Here we describe small moments from within research projects where apparent chaos was actually part of the process of doing the research – there was not necessarily a clear and concrete moment of discovery. What these moments signify might be 'nothing' (Truman et al, 2021), but equally to look at them might be to pay attention to how things emerge from a moment-by-moment unfolding process.

> Kate is sitting in the classroom. The windows are blacked out. A child is trying to film through a piece of fabric to create a particular effect. It is totally dark and some of the children (not all) are screaming loudly to produce a scary soundtrack. The child then balances the camera on a pedestal, and using a beam of light from a projector, films the look of

the light through the material. The effect was of a number of pinpoints of light. A film is then produced and shown with the title 'scary'.

A young person is jamming on a piano; loosely and out of the picture of a film camera, another young person is dancing/commenting. Two young people sway to the music nearby. Not anything obvious going on.

In these moments we can identify a space where young people's multimodal productions are taken seriously by young people, and are not particularly affected by performative adult-oriented viewpoints. They are regarded and given attention, rather than dismissed or ignored. They exist in their own right. This led to a flattening of hierarchies when doing research. We sought to uphold this flattening whenever we could, in our research practices and representational work (even as we similarly sought to engage in a sort of 'unflattening' [Sousanis, 2015] of complexities of self and personhood that sometimes are left unrendered in research).

In our writing about the projects, we acknowledge their mess and complexity (Law, 2004). Mess as a concept signals a kind of deviation, a 'strategic scruffiness' (Phillips and Kara, 2021, p 177) that maintains a dual focus on process and production. The language that we used to talk about the work was often located in the everyday. For example, we might talk about the 'stuff' of projects (Miller, 2009) as a word that signals this messy and grounded way of working. It is a way to avoid using terms that are more laden with theory, such as 'objects' or 'assemblages' or 'networks'. The word 'stuff' holds back the weight of theory for us. 'Stuff' is a word that works better in conversation than in written text, as it is often accompanied by gestures that serve to convey the bigness or smallness of stuff, the interstitial spaces that stuff inhabits, the relationship of one type of stuff to another (Goodwin, 1994). We also recognize the importance of dialogue, and it is within this genre of communication that we enter into ordering the 'stuff' of this chapter through our own experiences of theory and how theory helped us think.

Examples of mess could be the underbelly of meaning making in homes: the paper stuffed under the mattress or hidden out of sight (Pahl, 2002). It could be the writing in dust on a car, or the graffiti in a community park. Mess could signify the end of the classroom day where 'tidy up time' signals order and going home. Mess is found in the spontaneous bursts into laughter at seemingly inappropriate moments, or the literal mess created when materials used for a project are strewn everywhere in a shared space. It could include things that are magical, matter out of place and strange events, ghosts or happenings that can't be explained, objects that have their own agency beyond the everyday. Some practitioners (artists, youth workers) embrace mess as an ingredient for success; others (teachers, academics), possibly rightly, try and make things more coherent, in order to clarify the world. We tend to the messy end of things.

We talk throughout this book about the material conditions of how our work unfolded and the messy quality of those conditions. We accept that order can sometimes be useful in providing a clear structure within which to work. However, messiness has a quality that allows digression and diversions. This indeterminate messiness can provide the conditions from which a project could emerge. It is important to note here that missing from these projects was an adherence to strict definitions or boundaries about what was allowable, such that we worked to ensure that theoretical or methodological frames did not inhibit the growth of practices, material engagements and other outputs simply because they fell outside of expected boundaries. But mess is not (necessarily) unfettered chaos, as ruptures in shared practices were also attended to instead of being allowed to fester in destructive ways. Conflict, however, is not anathema to this ethos and, in fact, was viewed as vital materiality through which to enact our projects.

In many instances, the projects grew in the direction of these seeming digressions and diversions and non-linear narratives. In a sense, messiness guided the inner logic of our projects. In the 'Odd' project, for example (see list of projects in Appendix), embracing the idea of *un*planning was vital to be able to see what *was* happening, rather than remaining frozen in a place of worry about what wasn't happening according to plan. While planning with the teachers was important, we needed to stay within where the children were going and attend to their process of thinking, which required an element of mess, within the structure of school. Being 'off-task' was, in many ways, the point of the project – to be 'on-task' would limit the scope of what could be and what could emerge from the messy chaos of making together.[2]

In advocating for the practice of hearing children out, Haeny Yoon and Tran Templeton, like Paley (1986) before them, acknowledge the structural legacies of schooling into which children and teachers are subsumed: 'As the gatekeeper (Jackson, 1990), the teacher modulates noise levels to account for "learning," even at the expense of children's desires to do what they know best – play' (Yoon and Templeton, 2019, p 56). Similarly, researchers in these spaces can also reinforce, intentionally or inadvertently, these structures; or they can interrupt this inertia and take care to observe 'how children interpret and voice their social worlds, which can lend us insights into understandings about children and childhoods often obscured by our assumptions and desires' (Yoon and Templeton, 2019, p 57). Doing so is often a messy business.

---

[2] Thank you to Sarah Collins, researcher on the 'Odd' project team, for this insight.

Temporality

Messiness revealed the significance of context in these projects, specifically the ways that context is co-constructed, interrupted and remade through interactions between community members – some of whom became involved in our projects – and research team members, and a range of other interlocutors who entered into and out of the project space. Much of this contextual messiness and unpredictability was predicated upon the timing. Time spread further than usual, and things took time or happened in an instant, during a sunny afternoon or over the course of three years.

Collaboration depends, in large part, on the temporal context in which it unfolds. That is to say, the events and activities that surround the start of a project, be they a new policy that was enacted or the political climate that bends towards or against an issue, can shape what is doable within a project. That space, however, can also be sacred, and changes to that space can feel like a threat.

Given that we are writing this book in the wake of some projects that have concluded, while others are continuing and still others are emerging, we have taken a retrospective stance, which has allowed us to see how ideas travelled across the duration of our projects. Jay Lemke (2000) and Stanton Wortham and colleagues (2020) have outlined the significance of timescales in making sense of interactions and how they unfold. In an ethnography of a Latino community in Pennsylvania, Wortham and colleagues (2020) observed patterns of immigration over time from 1990 to the early 2010s, noting that the population had grown from 100 Mexican residents in 1990 to 35,000 in 2012, or from less than 1 per cent to over 22 per cent. Their ethnographic research into immigration in this town also elucidated the significance of timing for shaping the new immigrants' experience.

Elsewhere, Wortham and Catherine Rhodes (2012) offer a spatial conceptualization of timescales:

> A timescale is the characteristic spatiotemporal envelope within which a process happens. For instance, the emergence and development of capitalism, a process that in some respects has taken millennia, and in other respects centuries (Postone, 1993), is occurring across a very long timescale. In contrast, individuals develop their capacities and live their lives at ontogenetic timescales, across decades, drawing on but also developing sometimes-unique versions of more widely circulating models and categories. (p 84)

Wortham and Rhodes (2012) here remind us that timescales are elastic and are made within pre-existing categories but can stretch to recognize moments of encounter and relational ties. The immigration experiences of an adolescent in 1970, 1990 and 2010 may share similar characteristics by virtue of the stage of life and the nature of the experience. There will also be differences stemming from variation in sociopolitical contexts at the individual and group levels – for instance, being among the first to arrive in a new town in contrast with being the second or third generation to do so. This ontogenetic variation affects how a phenomenon unfolds, in addition to the expected impacts of social, material, cultural and spatial forces.

As such, we recognize that the timing of our respective collaborations with community organizations and schools intersected with the broader sociopolitical context, including our own ontogenetic development as researchers, the priorities and commitments of our participants, and the state of the community spaces where we located our projects. We attended to the rhythms and vicissitudes of everyday life in the projects we worked on, which did not necessarily tap into or align with the rhythms of universities in their requirements for semesters, terms and years.

Examples of timescales included the times in, of and around school – dinner time, home time and tidy-up time. Time can 'thicken' and slow while waiting for a fish to bite. Summer and winter create timescales of darkness and light, and the earth reveals slow timescales with remnants of past peoples. Finding histories that had not yet been written led to valuing the importance of oral histories. Community projects and teams attended to birthdays, weddings and funerals as part of the lifetime of a project (Campbell and Lassiter, 2014). Academic time, with its semesters, its lecture hours and its moments of encounter with students, has a different rhythm from the home and the school. In these in-between times, we lived the projects.

Spatiality

Being in a space, as described in the opening vignettes to this chapter, calls forth a story. The stories of the projects emerged from the sites and spaces from which they were formed and were developed from the logic of the site, whether this was a playground, a school classroom or a youth club. Places breathe with histories and memories of their previous inhabitants. Stones on the street, trampled on by feet rushing to catch a train, also once bore the weight of wagon wheels carrying whole families from one place to another; playgrounds are imbued with traces of the flora and fauna that once called the space home.

When we made a list of important places where our work took place, it looked like this:

1. the park
2. the pond
3. the street

4. the playground
5. the home
6. galleries and museums
7. the school
8. the court
9. the hill
10. reclaimed industrial space
11. the village hall
12. the pub
13. the library
14. the cemetery
15. the shop
16. the restaurant or cafe
17. the subway

Here, a picture emerges in the mind's eye of something located, within communities, within homes, leisure, life in the everyday. We have omitted the university, as it didn't immediately occur to us – the university can become the 'imagined other' as life is lived in these spaces (Pahl, 2016). The projects took their ideas from the space of the everyday.

We – that is, each of us in concert with our interlocutors and project collaborators – also made spaces, forged ways of being and interacting with the materiality of the physical places to where our work took us. By doing so, we could engage with the proposition that 'the world is an open process of mattering through which mattering itself acquires meaning and form through the realization of different agential possibilities' (Barad, 2006, p 141), attending to how objects, relationships, practices came to take up space and to have meaning that lingers beyond an initial moment of encounter or impact. Leander and colleagues (2010) describe 'spatiotemporal contours' with respect to children's mobilities as new technologies and digital media enter their lives, further commenting on the inseparability of space from time. Certainly the digital and non-digital (im)materiality of the spaces through which our research moved both affected and was effected by the ways that we (that is, university-based researchers and collaborators who occupied multiple social locations) inhabited these spatiotemporal contours differently in rhythm with the broader sociocultural forces at play at any given moment. Spatiotemporal contours are not merely where the mess of research lives and breathes; they *are* the research.

The idea of the university

The idea of the university as an actor or as a place of learning has been discussed variously, sometimes with a utopian hopefulness, but more often with a sense of despair at the ways in which universities have become chained to rhetorics of neoliberalism and advanced capitalism (Hall and Bowles, 2016; Amsler, 2019). Anne Carson (1999) proposed the idea of the university as something not rigid, but a network, which opens up a way of recognizing innovation. Moments within the university – moments of discussion, illumination, dissensus and critique – become important in their own right. The possibilities of non-hierarchical and extended relationships that stretch beyond the university, and are valuable not for any instrumental reason but for their own sake, then open up. This is the vision of the university that we work with here.

From here, we come to the idea of the university as a commons (Gilbert, 2013). In our joint article on the idea of co-production as a utopian concept, based in the 'not-yet', we (Bell and Pahl, 2018) drew on Jeremy Gilbert's idea of the commons, a non-hierarchical space of sociality, collectivity and subjectivity. One of the aspects of the projects described in this book is that they were variously utopian. They had titles like 'Reimagining Futures' or 'Imagine'. They aimed to resituate power relations by distributing resources and roles differently from the way in which they are distributed in universities, and often worked as a collective rather than a hierarchical structure. The projects operated both within and outside the university, in the 'cramped space' (Bell and Pahl, 2018) of co-production. However, we argued in this article that it is important to see 'beyond' this cramped space and articulate for something better. All of us, in different ways, have tried to change the ways our universities operate, from Lalitha's work to expand multimodal scholarship and digital futures, Kate's work on co-production and community, and Richard's work within teaching and learning; for us, this book represents an attempt to articulate the space of 'beyond' and the 'not-yet' of the university. As we said in our article, '[w]e should fight for academia as a space in which to co-produce' (Bell and Pahl, 2018, p 10).

Our final point is that the university/community dialogue is relational. In an article (Pahl, 2016), Kate argued that the university is an 'imagined other', a partner in a continued dialogue with itself, and in the projects, the university and the community often engaged in an imaginary dance between sites, spaces and thinking. This imaginary dance is relationally understood and is lived through meetings in coffee shops, through exchanges of writing, thinking and, more recently, Zoom calls where books were shared over a camera and discussions were carried out across time zones and in different stages of a pandemic. The university was enacted and relationally understood in those spaces. Each participant in the projects imagined the university differently – for some, it was a source of employment; for others, a place to learn and to get a doctorate; and for others still, it represented an irritating and somewhat confusing set of practices that got in the way of more important activities, such as teaching or social work or enjoying life or making art, music or fishing tackle.

In our writing of this book, we have tried to draw on ideas or words that are located outside the university as well as inside it. Our nomenclature included words or titles that young people arrived at such as 'Reimagining Futures' or 'Language as Talisman'. The language of the projects echoed our stance towards the university. Sometimes, university life becomes difficult. As we strive to make our universities 'better' in some way, we are also conscious of the many institutional failures of the university. Some of our collaborators outside the university have been involved in doctoral study. Many describe their status as both insider and outsider; as Milton Brown and colleagues (2020) describe this position, they are 'standing in the gap'. The 'I' can become the 'we' in a really positive way in a university/community project, but the university can also block our desires for things to be better. We all wrestle with this, but continue to be hopeful.

Research methods and universities: the pedagogization of research

In this book, we explore the ways that universities do or do not support collaborative research. Part of the reasons for universities' difficulties with the idea of collaborative research lies in the way in which long-standing methods and practices support a particular rhetoric of what 'research' means (Law, 2004). That is, research can signify an intentional distance between the researcher and those who are researched, where the seat of power is firmly rooted with the former, as an agent of the university.

Certainly, we see this now-archaic model of research changing. Methods of enquiry (research methods) have become increasingly specialized as new tools and processes of observation, analysis and representation have been developed and continue to be refined. And courses about decolonizing methodologies, multimodal discourse analysis, participatory ethnography and other methodological topics that close the researcher–researched distance continue to proliferate in university course catalogues. Even still, the prevailing notion is one that research methods are simply things to be 'taught' and received in courses. Here, we call that idea into question and explore it in more detail with respect to the ideas of Brian Street, an anthropologist of literacy.

Brian Street (1995), in his book *Social Literacies: Critical Approaches to Literacy Development, Ethnography, and Education*, introduces the phrase 'pedagogization of literacy' to refer to the ways that the multiplicity of literacies is too often reduced to a singular definition that aligns with literacy rules that are taught in school. This institutionalization of a practice that is situated, quotidian and deeply human – that is, the practice of composing and conveying meanings using available resources, including reading, writing and speaking – Street argues, is essentially an act of reduction and constraint in the name of neutrality and standardization.

One might argue that a similar set of constraints is evident in university-based research methods courses in which new generations of researchers receive some of their training for pursuing research in communities. Research has, in effect, been pedagogized in and by universities, the result of which is a spate of studies that produce staid results or perpetuate possibly harmful conclusions about communities. Having taught methods courses ourselves, we are moved to ask whether there is room in university research training for the sort of messiness and the recognition of temporal and spatial fickleness that we mentioned earlier. We wonder how the researcher might embrace a sense of methodological consistency and commitment without becoming fixated on a distanced notion of rigour that can threaten to undermine the very nature of research. Here, consistency is not synonymous with uniformity; rather, consistency is understood as honouring the relationships on which collaborative research is built.

In this spirit, we consider the forms of research: how collaboration is meted out in relational practices, artefacts of documentation and representational forms that travel beyond the immediate boundaries of a project. The contours of these dimensions of form are further nuanced in the chapters of the book, but we take a moment here to set the stage for you, our readers, to (re)consider form with us. We wonder about opening up both the modes and methods of inscription of research encounters beyond one form. We have ourselves worked with film and multimodal forms (such as the hypertext, the collage, the zine) as a way of expressing research ideas. We have experimented with the idea of 'queering the form' as a way to acknowledge difference and speculative accounts of realism that engage with fictional realities and new, un-met experiences (Shannon and Truman, 2020). We have started to imagine knowledge as being untethered from universities.

The (un)pedagogization of form

Picture this scene: a global pandemic. By some estimates, nearly a billion children experienced some form of educational displacement resulting from school closures. Norms of how and where learning happened were upended, and in the wake of these disturbances emerged initially a handful, then countless more, spontaneous moments of creative production.

In one such moment, Mo Willems, the children's book author and illustrator, and Yo-Yo Ma, the celebrated cellist, came together during the early months of the pandemic to collaborate and co-produce art. Ma played pieces of music with his cello while Willems doodled with pencils and markers on large sheets of paper. They met like this (experiments, they called them) a handful of times and then compiled the meetups into a 25-minute video that was posted to the Kennedy Center's YouTube channel and dubbed the 'Yo-Yo Mo Show',[3] in a portmanteau of their names.

After one of the experiments, Willems held up the drawing he had completed while Ma played an excerpt from Bach – Bourree 1 and 2 from cello suite no. 3 – and reflected on what he had produced: three jelly-bean shaped blobs that had been marked up to resemble birds. "Once I had the forms, I was really kind of dancing around a little," Willems says in the video.

In another experiment, Willems and Ma discussed exercises each does in their respective artistic practices before they embarked together on an improvisation. Willems drew a long line on a piece of paper stretched across the length of a long table and the line served as anchor and muse for the subsequent drawing. Ma, seated behind his instrument and watching his friend on a screen in front of him, started to play his cello in rhythmic accompaniment and response to the drawing that unfolded before his eyes.

At the end of this experiment, which lasts just minutes, Ma attempts to explain what transpired from his perspective: "I'm just looking at the thing that's actually activating something in me that automatically is making something come out. I'm not thinking any more. It just starts to happen."

Willems' and Ma's reflections are redolent of Jane Hirshfield's (1997) observations about musicians and dancers:

> Violinists practicing scales and dancers repeating the same movement over decades are not simply warming up or mechanically training their muscles. They are learning how to attend unswervingly, moment by moment, to themselves and their art; learning to come into steady presence, free from the distractions of interest or boredom. (p 4)

These two artists 'attend unswervingly' to their art and 'come into steady presence' with one another. The hairs on a bow move across strings to produce melodic riffs and a blank sheet of paper is transformed into creatures and worlds with colours and shapes. Form functions as both purpose and method.

The modes of inscription – Ma's fingers moving along the neck of the cello, delicately dancing across the fingerboard to coax music with his bow from the wooden instrument, and Willems deftly wielding coloured pencils and markers to make marks on large sheets of white and brown paper to produce recognizable and otherworldly images – exist in relation to one another, and are very much *of* the space out of which they emerge. The form of their collaboration was more than the sum of its parts, which is only

---

[3]   The full video can be found on the Kennedy Center's YouTube channel: www.youtube.com/watch?v=vEh9vsB9OfE.

partially captured by either the drawings or the musical compositions, and is wholly incomplete without the surrounding context of a distant audience tuning in from around the world to watch this interaction unfold.

We think about form as resting both within and rising out of the methods by which modes of inscription are utilized in a given situation. Those modes are not predetermined, even if the scope of a given research project is defined: a study of collaborative writing on digital platforms, enquiry into the role of drama in community building, a project to support the development of zines as part of a poetry library. In each example, drawn from real life experiences, the form that was initially at the centre of enquiry did not extinguish the sparks of engagement between participants in our projects and the broader materiality of the spaces through which the work moved.

Collaborative research projects, which gather together people who hold varying degrees of connection to a central organizing question or focus, often have an outside logic that emerges from what is happening in the world. Texts find their ways into project meetings or interviews as young people bring music into the conversation or share observations about recent popular cultural texts or events. These project rhythms do not necessarily fit with the neatness suggested by proposed research designs; in contrast, the rhythms of projects are messy, temporally impacted and materially emergent. Likewise, the ways that people write themselves into the work of the projects – whether they are repurposing available tropes and labels or remixing existing methods like fieldnotes with unexpected notations or format – created an often wider variety of research outputs available for sense-making about a phenomenon (see Jungnickel, 2020).

These shifts impact how the stories about collaborative research can be crafted and disseminated. In other words, the representational form can be altered by the interactional practices and communicative forms that unfold in a space. We say 'can be', because the researcher must be willing to 'attend unswervingly' (Hirshfield, 1997) to the emergent forms, even as they contrast with, contradict or, in some instances, violate the institutional norms and expectations that externally frame a project.

Collaborative and co-produced research is attempting to open up both form and content, specifically by creating opportunities to attend to the nuanced knowledge that is found on the edges of a pond, on the streets, in homes and with communities that are too often kept at a far remove from the knowledge privileged in the university. This kind of work requires shifts in orientation, methodology, and practices of knowing and representation that are necessitated by the long-held exclusion of community voices and knowledge from university canons (Rasool, 2017).

While universities bear the weight of responsibility for providing the infrastructure necessary so that a vast range of activities can gestate and emerge, they must not function as impenetrable fortresses whose internal logics are immune to change. We write in this book about our practices of creating the conditions of researching 'with' communities in order to further sustain collaboration and partnership. By doing so, we urge researchers to regard universities and the pedagogized forms of research and knowledge production critically, and cultivate openness in service of attending to spaces from which knowledge can come forth unbidden.

Institutional blasphemy: an argument for co-production

We undertake this work as researchers and co-enquirers in the communities where we locate our work, with the people whose experiences are the driving force for the praxis that unfolds as a result of collaboration. Community organizations, therefore, hold the roots of the collaborative work and practices that we detail in this book. Borne out of conversations, a chance encounter, mutual acquaintance and other such unplanned and largely serendipitous moments, collaborative work can begin to take root. For such work to be co-production requires additional tending to the norms and practices of research. Often, this tending also necessitates rethinking ways of doing research that have been situated in relation to the idea of 'knowing' into a wider, diverse, polysemic space (de Sousa Santos and Meneses, 2019, p 118).

In short, we propose to engage in a sort of blasphemy. We do so for purposes of widening the scope and function of research to not merely pay lip-service to the inclusion of participants' voices in shaping knowledge, but to have their ways of knowing and being be centrally transformative of the research endeavour. This can unfold as messy, unruly experiences in which failure and unpredictability are the norm (Thomas-Hughes, 2018).

This blasphemy was not without cost, for there were many; nor were the kinds of wide open participatory efforts that we discuss in this book free from the politics of research. In the early 2000s, publishing venues for articles that contained animations or other media were not plentiful, and when they existed – often as 'online journals' – their status in university-based metrics of academic performance (tenure, promotion) was tenuous. Thankfully, as the expanse of digital and portable technologies has grown, and particularly as the communities with whom we collaborated deepened their interest in expanded modes of communication and representation, more academic journals have been welcoming of multimodal scholarship. Likewise, other venues for disseminating research – blogs, social media, virtual convenings, exhibitions, immersive experiences – also serve to retain the generative complexity of the research and lower barriers to participation in the representational work of research.

A further cost, or challenge, lies in the fact that research grounded in the sort of messy collaboration and co-production that we pursued took time – time to develop rapport, earn trust and credibility, cultivate emic understandings – and for each of us, this work unfolded alongside a full slate of other responsibilities like teaching, doctoral advisement, publishing expectations and grant writing. Thus, an already meandering path of enquiry was further complicated by these layered responsibilities and the roles they implied. To say the benefits outweighed the challenges would be a gross oversimplification; the challenges (of time, resources, relationships, welcoming venues) also became the work, became sites of enquiry, change and transformation.

In these pages, we contend with what those concepts have looked like in projects around the world in which we have been engaged. We conclude here with an overview of the projects that anchor our discussions throughout this book. These are not extensive descriptions of methods or sites, but rather a conceptual overview of what types of projects these were, and the broader ethos and logics that guided their evolutions and that continue to guide our enquiries into them today. In Chapter 2, we delve more deeply into the theoretical commitments that grounded these projects and describe some of the thinking behind the work.

Connected communities programme: funding reimagined

The de-pedagogization of research as a project was helped in the UK by a funding programme that lasted from 2010 to 2022 called 'Connected Communities'. It was initiated by the Arts and Humanities Research Council (AHRC) in the UK in 2010 and, at the time, there was a concern that it was connected to a conservative agenda that meant it validated austerity over state interventions. It seemed to respond to a turn within the UK political landscape to a focus on communities as actors, and a rejection of the idea of the monolithic state – an ideological approach described as 'The Big Society' by the then Conservative Prime Minister, David Cameron. However, there were some distinctive aspects to this programme of research. Key to its principles was the idea of research proposals as being developed *with* communities, not on communities. The focus was on the ways in which arts and humanities research, which sometimes appeared to be distant from the mundane aspects of everyday life, civic participation and social relations, could be useful in rethinking concepts like community, civic engagement, participation and the regulations of engagement. The projects that were initiated were often co-produced, with community partners, and took their logic from community concerns and deliberations.

One of the legacies of the 'Connected Communities' programme was a set of methodological approaches that privileged dialogue, dissensus, polyphony and reflective practice. These qualities are discussed more fully in Chapter 2. The approach taken resembled the classical idea of *skholè* from which we get the term 'school' – this is here discussed by one of the key reports from the 'Connected Communities' programme, called 'Creating Living Knowledge':

> The sorts of spaces, times and practices that are being developed by the community and academic partners in these projects, in other words, bear a closer resemblance to an old Aristotelian idea of 'skholè' than they do to the fantasies of management consultants eyeing up the best ways to make universities accountable. Skholè, the word that gives most European languages the word school means, among other things, a time of freedom, a moment of reflection that is an important part of the rhythm of living. It is the space within which to reflect upon progress achieved, to re-examine core purposes and values, and to experiment with trying out alternatives. Skholè is the site in and through which both action and theory are developed through dialogue. It is a time in which the different focus of knowledge held by individuals and organisations are released from their habitual associations and made public, available for common use. (Facer and Enright, 2016, p 152)

Institutions need to be 'intimately connected with the world' (Facer and Enright, 2016, p 152) to make sense of the world, the report argues. The idea of the enmeshed, entangled spaces of thinking that this kind of work produces is contrasted with our description of the three spaces outlined at the start of this chapter. In Keri Facer and Bryony Enright's (2016) report, the conundrums of this programme come to the fore. Here they describe the 'Connected Communities' programme: 'Since 2010, the programme has funded over 300 projects, bringing together over 700 academics and over 500 collaborating organizations on topics ranging from festivals to community food, from everyday creativity to care homes, from hyperlocal journalism to community energy' (p 1). This aim was laudable, but came with various caveats. Often, in the early stages, academics and community partners felt their way, and had to deal with the complexities of partnership, including struggles with such mundane issues as payment, time, space and commitment from universities to communities to do long-term work rather than short-term funded projects. This work also produced different kinds of knowledge that could be understood differently. The process of doing the projects remade both knowledge, identities, and ways of making and thinking about things in the doing of the projects. Working on the projects offered different roles, Facer and Enright (2016) argued, including those of catalyser, integrator, designer, broker, facilitator, project manager, diplomat, scholar, conscience, data gatherer, nurturer, loud hailer and accountant. In these roles, academics sometimes felt that they were 'de-skilled' and lost their claim to authority and knowledge. Indeed, sometimes, we (Richard and Kate) felt as if the community 'knew' more than we did, when sitting by the pond watching young people fish much better than us. There was a sense that the 'Connected Communities' programme de-skilled academics from their day job, and replaced this with a much more complex set of tasks.

In the US, while research collaborations between academic researchers and communities and community-based organizations abound, funding to support collaborative partnerships has been more limited than funding for empirical research that maintains the boundaries of a researcher–researched arrangement. Research-practice partnerships is an umbrella term that connects 'research, policy, practice, and community work', in an effort to 'address persistent challenges and systemic inequities in our schools and communities' (Farrell et al, 2021, p iv). Funding sources in the US (for example, the William T. Grant Foundation and Spencer Foundation, as well as the Institute for Education Sciences) developed calls for proposals that promote 'long-term collaboration aimed at educational improvement or equitable transformation through engagement with research' (Farrell et al, 2021, p iv). In this book, we take on the idea of long-term collaboration with a focus on co-produced research. Two examples of this shared commitment that we have sought in our collaborative research partnerships are expanding participation through the intentional engagement of multimodal practices and cultivating audiences for research insights that go beyond the common sites of publication and authoring. These varied outputs call for funding agencies to be open to descriptions of outcomes that may not have a readily discernible taxonomy at first. Grants for making research accessible to the public, for instance, would be made even more effective if funding for youth media creation was included as part of dissemination costs. The unpredictable ways that young people, who might be the focus of many studies, can bring their curiosities to bear in creating artefacts to circulate to audiences that might not be on researchers' minds may prove valuable for the research as well as for the young creators.

These kinds of projects tended to take place across interdisciplinary groups of people. Activity would then be focused within a site. While there were meetings within the university (see the description of a university event at the beginning of this chapter) the university was not necessarily the site of the knowledge investigations. What this, therefore, required was a way of working that involved thinking in groups. There are a number of different approaches or methodologies associated with the idea of thinking-in-groups. One of the most popular, and useful, is the concept of 'action research', and its methodological cousin, 'Participatory Action Research (PAR)' (Kindon et al, 2010). This way of working tends to refer to the action research 'circle' as its orienting framework. The circle might begin with defining an action to be taken. The action then is focused upon, but, importantly, a period of reflection is used to evaluate and refine that action and change the approach. The cycle then begins again.

PAR expands on action research by actively building in activities and methods that broaden participation beyond a bounded or predetermined research team and, importantly, this broadened participation can shape and change the direction of the research. In this way, PAR goes beyond paying lip-service to principles of inclusion and creates sustained opportunities for members of the community to assume an ownership role over the processes and purposes of the research. The action is defined, but the group can be broad and can include people with a variety of different experiences. The idea of the action research circle provides a structure where this activity of reflection and action takes place, with expertise equally valued across a range of participating researchers from inside as well as outside universities (Kindon et al, 2010). Importantly, PAR is a methodology that values practitioner knowledge and that can be carried out outside of the university. Its ways of working are not mysterious to non-academic researchers.

One of the most helpful aspects of PAR is that it provides a structure for listening. Experience becomes the touchstone for learning and discussion. Sarah Banks et al (2014) describe how they used a form of 'collaborative reflexivity' (p 37) to describe the process of thinking together. Drawing on a co-enquiry model developed by John Heron and Peter Reason (2008), this approach provided a way of working that valued group experience and knowledge in a respectful and participatory way. Researchers could be both participants and subjects, drawing on their experience to come to shared understandings. This approach required a radical epistemological stance of valuing community knowledge.

When thinking about how universities and communities can work together, Angie Hart and David Wolff (2006) identified a 'communities of practice' approach, drawn from the work of Etienne Wenger, to be useful. A communities-of-practice approach recognizes the ways in which a diverse group of people can work together through shared practice and enthusiasm and how this group work can be supported (Wenger, 1998). The key ingredients for this way of working include an awareness that hierarchical ways of doing things are not always successful in creating change and a real desire to work together outside of managerial systems. Examples of trying to mitigate that can include providing community partners with library access and desk space, as well as university resources more broadly – for example, access to rooms and equipment – and, thereby, blurring the boundaries between university and community in concrete ways.

The idea of 'collaborative ethnography' arose from an ethical concern around the representation of communities and a resulting questioning of the authorial rights of the ethnographer to describe communities without actively involving them in interpretative work and the writing of the text. A search for a more reciprocal and ethically charged practice resulted in a shift in thinking around whose voice counts and why (Lassiter, 2005). Collaborative ethnography as a mode of doing research requires the willingness to allow an ethnographic text to 'unfold in unexpected directions – where intellectual and moral exchange not only deepen interpretation, but more completely extend the dialogic metaphor to its political and ethical implications' (Lassiter, 2005, p 13). This idea of unfolding in 'unexpected directions' is what this book is intended to explore. Instead of one voice describing an encounter with a field, the resulting texts from a collaborative ethnography are multi-voiced and dialogic, with a reciprocal mode of analysis built into the structure of the research (Campbell and Lassiter, 2010). This raises ethical and moral questions as to whose voices count and where and when.

Another approach often used in the community/university projects included a particular focus on embodied knowledge (Barrett and Bolt, 2010). This was knowledge felt and experienced through handling and making things together. In working with communities in situated ways, relational modes of creating things together became useful. This work touched on relational aesthetics and socially engaged art as modes of thinking and relating to each other (Kester, 2013). Steve Pool (2018) articulated the various ways in which artistic methods have been used within community projects. Pool identified a hybrid mode of knowledge production that spanned research methods, art forms and artistic practice that often could be found within collaborative interdisciplinary projects, suggesting that everyone can 'have a go'. In such projects we found that no single approach or method was the answer to the range of circumstances and opportunities that surfaced. Marjorie Orellana (2019) writes about the impossibility of fully capturing all that happens in life with words on a page: '[b]y acknowledging the sense of overwhelm, we may be more able to see how paltry our efforts to see it all are – and then *use* our initial impressions to think more about what is salient to us, and why – and open ourselves a little more to other possibilities' (p 51).

Conclusion

This work is about collaboration and stories. It is also about relationships between universities and the communities that they work with. Using the concept of 'poetics' has opened out a different register for 'knowing with'. We had to change, protean-like, to adjust to the projects. Our orientation is to become more hopeful in the way we work. The book is an exercise in forward thinking and emergence, as the title suggests. The book also tries to unseat authority through the stances developed within the projects. The book asks researchers to become vulnerable in the process of doing research (see Singh, 2018). Research becomes reframed as being about the encounter, one that cannot necessarily be predetermined. This encounter can be understood as messy, open and less clear, but devised and developed in the moment. This is a polyphonic space, which speaks with many voices.

Recognizing messiness as a condition of the work ensures it flourishes in new ways. This means that time as well as space is stretched to recognize the time frames of the projects. The importance of *emergence* as a condition in which to do research is part of this vision, which also involves locating and describing the context of the work as an integral part of the research. Holding an 'as if' speculative vision is important for this type of work.

In this book we have begun to map out the possibilities for a new language of description for the work we do. Our aim is to provide, through a series of chapters, interludes and provocative explorations (worldizing, worthiness, enchantment, embodiment, hypertext, unplanning) a set of ideas about collaborative research that can, ultimately, try to reshape the language of engaged research with young people and communities.

Collaborative Questioning

Kristine Rodriguez Kerr

The questions that follow were taken (as is) from collaborative research team spaces from 2009 to 2019 that spanned blogs, Google Docs, emails and shared workshop/fieldnotes. More specifically, each question was crafted by a research team member and then shared, in some way, with the rest of the team, often without any expectation of a direct answer. Instead of answers, most of these questions were meant to push individuals and the group to think, consider and engage with our site, participants and our deepening understandings anew. At any given time, the research team consisted of our principal investigator, youth interns and graduate students, with the latter serving as project coordinators, workshop facilitators and youth mentors. Each of the questions included has been shared without attribution and could have been authored by any of these team members. Some individuals rotated in and out of the project, and others continued with the work across our decade of data collection.

How do you reimagine your future?

What happened after our workshop ended? What happened before we arrived?

This breaks my heart/makes me angry/hurts my brain – how is this possible?

Where do young people develop the strength to follow their own path? Who follows? Who leads in early adolescence (for example, 10–14 years old)?

What kinds of support do early adolescents want? What are their everyday goals?

What do we hope will result from this project?

Are we meeting today?

How should we keep moving forward?

I'm moved by both of your reflections about not always taking or being steered down the 'right' path. What contributes to this?

What will we need to do to make this workshop possible?

How are we bringing to light what possibilities and opportunities these adolescents may have even after having encountered one mistake that may have already shifted their mentality into believing that 'a good future' is impossible?

Peace and quiet: we all need that kind of time here and there, right?

What do you/participants want to know more about?

I asked "Why are you here?" She answered, "Is this a trick question?" Why would this be a trick question?

They [participants] are making big decisions, almost daily. Who/where/how are they supported with these big decisions?

How can we plan for participants coming in and out?

If [he] stops by again someday soon, I'd like to see if I can talk with him and if I can find out what is 'the it' that made him come back?

What changes have happened with the opening of the new location?

How can we move from isolated use of equipment to 'this is what we'll do a project about' and where will it be shown? (How can we make workshops more meaningful than one-day events?)

What are our expectations for the project this year?

How can we continue to build workshops that come from open-ended expectations?

I wonder why [he] switched his story? I asked him if it happened in school and if he had gotten in trouble?

Is it even possible to not trash talk/boast when playing chess?

I found some of the questions hard to answer and wondered why people don't ask these questions more?

Do they share this much in school?

It says 'establish a relationship and get to know your students/participants' ... HOW?

Where/when did you feel listened too? What made that possible?

[She] said she learned in school and she played by herself – online, I wonder?

After time passed I asked myself "how come I'm sitting here across from this kid, and not talking?"

What happened afterwards?

But let me ask y'all a question: was it really worth it?

What was the prompt again?

Who determines what counts as a 'punch'?

So, no meeting today?

What is the role of facilitators in that moment AND how to make sure the workshop makes room for/within those moments?

Can we print a big print of the group collage?

What types of pedagogical insights are we having?

What are additional things we can imagine doing?

In our embodied multimodal pedagogy (space to be) what ways of being are not easily identifiable?

How do we relearn to look for these moments?

It's fun. It's goofy. It's play. How do we take that seriously?

What are the things that don't get talked about?

How do you enter a research space? Physically, relationally, technology, corporeally, in which subsequent knowledge is produced about that space and the people in it?

How does/can a piece of media embody relationships?

What is being observed?

How do we know what we know?

I wonder – is it too much too soon? Should we take one concept and stretch it over all the weeks? Or not?

How do youth employ multimodal tools to craft different narratives about themselves OR cultivate different narrative practices?

What does it mean to KNOW something?

Are we knowing them better? Or knowing them differently?

How are you understanding 'wellbeing' here?

What does it mean to be a young person who is involved in a complicated system?

How do staff respond to participants' everyday lives?

How do we build familiarity and consistency when kids are only there for a short time and we are only there for a couple of days a week? Is it possible?

What have they enjoyed? What do they want more/less of?

I wonder, do our conversations with them stay with them when they go home? They stay with me.

What are the instances of authoring you see happening?

What do you think of these prompts: what do you like about school? Why? What do you dislike about school? What would you do to change it?

I wonder what triggered that idea?

This conversation led to another about lies, lying, and when is it ok or perhaps noble to lie?

Do you want Pepsi or Orange soda?

Maybe this was intentional, since this is a very active engagement?

I wonder how that came about?

Notebooks? Blogs? Cellphones?

Are there places for private thoughts in fieldnotes?

It's clear (to me) that modalities and digital spaces inform my teaching practice, but how is my use of them informed by my teaching practice?

Who are the audiences? And how to know what our audiences are assuming?

How are nurturing spaces created?

What's with this kid? (I'm so curious to know)

Why was it so empty? Where were all the other participants? Were they running late? Were they not scheduled to come in today?

I knew we had to pause the movie time and time again, but why was it only getting better?

I wonder if they learned about jazz from their parents, or neighbourhoods, or via schools, or from watching movies, etc.? They knew much more than I did.

What is it like to be involved in a complicated system (sometimes against your will)?

So how do we go about doing this? What are some things that have worked?

Why do we live in a world, in a society, where this even needs to be stated?

Any thoughts on the questions?

Poetics

Kate Pahl, Richard Steadman-Jones and Lalitha Vasudevan

The earliest instance of the term poetics according to the Oxford English Dictionary (OED) is in John Milton's 1664 work 'Of Education' where it means, 'the aspect of literary criticism that deals with poetry', or, 'the branch of knowledge that deals with poetry'. The term retained this meaning for centuries to come. However, the OED also indicates that in the 20th century the term was used with a broader meaning, 'the creative principles informing any literary or social or cultural construction', for example in Alan Sheridan's translation of Lacan's *Écrits* we find the sentence, 'this notion must be approached through its resonances in what I shall call the "poetics" of the Freudian corpus'. This expansion of the term in some ways constitutes a return to its original meaning in Greek. The adjective ποιητικός 'poiētikos' derives from the verb ποιέω 'poieō' to make or to do, the first meaning of which in Liddell and Scott's Lexicon is, 'capable of making, creative, productive'. However, in 20th century usage, the term in its broader sense is coloured by its association with verbal art. It does not suggest a routine procedure for production or manufacture, but a creative process involving imagination and judgement. It is in this sense that it appears in the title of Gaston Bachelard's book, *The Poetics of Space* (1958/1994) and Eduard Glissant's *The Poetics of Relation* (1997). The former explored the architecture of the house as a creative endeavour grounded in cultural memory and a certain kind of lyrical responsiveness, and in the latter the making of history is similarly understood as a creative process which constitutes a transformative intervention in the world itself. In using the term, we want to highlight the notion of collaborative research as a creative endeavour.

When we use the term 'poetics' of research, we are talking about a process of designing and cultivating the conditions for collaborative research to flourish. The reason for using the word 'poetics' rather than 'methodology' is that there is something artful about this process. It requires a thoughtful and empathetic response to the conditions as they develop (see also Mirra, 2021). In this sense, the endeavour of collaborative research is always a journey, always an ongoing and unfolding experience that calls on participants who occupy different roles – as researchers, students, teachers, community members, staff, artists – to be willing to be transported as a result of their participation. Much as Luis Moll, Norma Gonzáles and colleagues (Moll et al, 1992; González et al, 2005) have argued for local funds of knowledge to be recognized and integrated into classroom discourses and practices, research projects similarly create conditions and spaces, however permeable, that can also reflect wildly different tenors depending on how those projects' spaces are initiated, designed and cultivated. We recognize that scaffolding research efforts to be more inclusive of the 'cultural funds' that participants possess is not accomplished in a single step. Beyond inclusive practices alone, a poetic approach means that we sought to harness the knowledge, practices and commitments that all participants brought to a project.

Here, we introduce the reader to some of the key concepts in this book: (1) unplanning, (2) work, (3) story, (4) embodiment, (5) polyphony, (6) worthiness, (7) audiencing and (8) dis/enchantment. These concepts enable a set of insights to be built up about collaborative interdisciplinary research and constitute a poetics arising from that work. If the ideas of spatiality, temporality and messiness called forth in the previous chapter provided necessary anchors for grounding our thinking about collaboration and co-production, then the concepts that we introduce and explicate in this chapter are reflective of how our projects grew. The ideas discussed here are central to a poetics of research and can form the ground of thoughtful and empathetic responses to people and situations. These concepts illuminate the projects, but do not exactly define them methodologically. Rather, they express the orientation that we took towards the research. Our understanding of these concepts has crystallized over time; in some cases, for example, that of the concept of 'work', we became conscious of the idea while working on a particular project, but then realized that in a less conscious way, it had also informed earlier projects.

The concept of 'poetics' can help elucidate the ways in which a humanities approach to collaborative and co-produced research, centred on meanings and cultural concepts, can work. These concepts are embedded and realized in their fuller sense in the chapters that follow, and in our concluding chapter, we revisit them again. As concepts emerged during extensive conversations and reflections across our various projects, so, too, did they bring to light revelations about the nature and purpose of the collaborations with communities and organizations in which we had been immersed for decades.

Unplanning

Our projects have tended to value such phenomena as serendipity, improvisation, chance, indeterminacy and contingency – what we have termed 'unplanning' as an umbrella idea. In the book we explore the contours of unplanning in an effort to explicate both an appreciation for and commitment to retaining such a stance in our research. This could involve an awareness of the need to relinquish the kind of control that comes within a fixed methodology. For example, one could undertake a week-long programme of working with children without a strong sense of predetermined outcomes. This orientation requires a different kind of practice than one that is often held up as a goal in methods texts. It is quite rare for a social scientist, for example, to go into a field without a clear sense of direction, research questions, interview questions and focus group suggestions. We suggest that this way of working can open up new avenues as children and young people lead the way (Gallacher and Gallagher, 2008).

However, the concept of unplanning does not entail the complete rejection of planning. Instead, it involves a sensitivity to things as they unfold, and an awareness that a project might take off in a new direction – a kind of indeterminacy in the work. This then leads to new ways of doing things that become embedded in the practice of research. For example, ethical procedures have to be recursively worked through with children and young people. As well as asking for agreement at the start of a project, an ethical stance might emerge that required permissions along the way, as the work took directions that were new. Such an orientation requires preparation and scaffolding, but also a kind of nimbleness, which allows one to embrace the unpredictable. This way of working requires what we call 'structured unplanning' – a type of planning which allows for unforeseen developments and serendipity.

Work

We talk throughout the book of our 'work', a term that originated from art-school discourse, and that permeated many of our projects with the inclusion of artists as collaborators. A work is an object and a product, but work is also the process of making that gives rise to that object. Thus, in looking at a work (in the sense of a product), one can ask the playful question, 'Where is the work?' A response to this question clarifying the processes of making, doing and thinking would be an expected part of the art-school 'crit'[1] and

---

[1]   A crit is a formal group session where feedback is given to students by their course tutors and small groups of fellow students in the studios.

might reveal dimensions of the work that are not immediately apparent in the object itself. The concept of 'the work' gives a sensitivity to the actual process of making. It encourages us to bring our attention to process rather than simply product.

As an example, one of the contributors to this volume, Steve Pool, sent an email to Kate Pahl about a pirate ship that he made for an adventure playground. Steve writes:

> When we ask, 'where the work is' it is an open question that is meant to encourage the art student to think about what is agentive in relation to what they have made. For example, if I were in a crit about the pirate ship I built at the playground and somebody asked me where the work was, I would say: The work is not really in the object, although it carries a certain aesthetic and is in some ways a giant public sculpture. Its intention is to act as a symbolic object that stands for investment. I would not call it participatory or social and its making was very personal. It transcends the research project that partly funded it and most definitely afforded a situation where I had the time and resources to make it. I would say for the past ten years my practice has really focused on the potential to see artworks as lines of flight from an assemblage of social and material relations. The pirate ship as an object is 95 percent a piece of play equipment and 5 percent a work of art. So the idea of the work or the question of where it is is essentially a spatial one. It reminds me of asking teachers where the Curriculum is? is it in the store cupboard? where does it live? (Steve Pool, email, 4 August 2021)

In an open-ended project, which yields unpredictable and serendipitous outcomes, there is a value in constantly asking oneself 'where is the work, where does it live?' If we do this, we may discover that the 'work' in fact lies in apparently inconsequential details, traces of process which, given a traditional framing of research, one might not see as part of the 'work' at all.

Story

Stories are integral to an idea of the 'poetics' of collaboration because they are a central means by which knowledge is preserved and transmitted within communities. Stories are a way in which knowledge is made. Stories can be expressed within oral traditions, written texts and inscribed artefacts, objects that bear narrative weight. Stories *are* knowledge, passed down through generations and across cultural networks and institutions. Clint Smith (2021), for example, traces the stories of slavery across the history of the US, and in doing so, animates accounts of enslaved people and their descendants that remained hidden or protected for hundreds of years. He links stories together with remembrance, a sacred act of keeping alive the experiences of people who lived and endured unimaginable horrors. In his writing, Smith moves from human observations made while listening to an insightful tour guide at the Monticello plantation to the stories of family separation that haunt the halls of the building and ripple out across the lush green landscape. Such 'semiosis of remembrance' (Brockmeier, 2002) is based upon signs that, even as their meaning may evolve over time, hold transportive qualities.

Stories and narratives also drive the way that researchers communicate their findings. Scientists working in laboratories labour to find the connective tissue that links experimental data and hence convey a larger point through a coherent narrative. Economists reach out for common tropes to frame their quantitative analyses of patterns of consumption for an audience. Similarly, qualitative researchers, ethnographers, and education and humanities scholars draw on stories as an organizing principle for communicating the layered experiences in and of the communities and people who are central in the research. Thus, the 'poetics' of collaborative research involves the interaction of multiple stories, each with a different narrator and a different stance.

But stories are more than simply communicative devices that convey knowledge. They are also purposeful – a story is told to some end and for this reason it is important to ask, whose stories are being shared? Quieted? Exalted? Whose ends are served by the telling of particular stories? Stories, as we know, are 'important tool[s] for proclaiming ourselves as cultural beings' (Dyson and Genishi, 1994, p 4), and so, once again, the question is whose understanding is being propagated by the story. Who tells the story and how it is shaped is a key part of the poetics of research. We might say that a story is like the agar in which the seeds of knowing are cultured, and collaborative research depends on cultivating the conditions for shared storytelling and thus shared knowledge making.

Here, we talk about storying as a way of working, of being, of seeing. However, stories are not crafted verbally, alone. Dorothy Holland and colleagues (2001) assert the importance of articulating and crafting desired selves that, when regarded through a multimodal lens – that is, by attending to the multiple levels at which meaning is communicated – open up ever more sites for perceiving and enacting an ethos of storying. In her ethnography of an insurance company, Charlotte Linde (2009) enquires into the ways that stories function within and around institutions. She notes that the story of the institution is held in many ways, by both people and artefacts, which encounter, produce and are produced by it. Storying and restorying (Vasudevan and Rodriguez Kerr, 2012; Thomas and Stornaiuolo, 2016) is ongoing and active work that seeks to call attention to voices that may have been kept at a far remove and to create conditions for stories to be heard, made and shared. So, if one important question for researchers to ask themselves is, 'where is the work?', another is, 'what is the story?'

Embodiment

Our projects were embodied as well as imagined and hence the poetics of research must extend beyond the linguistic to include embodiment as a key aspect of the work. Indeed, 'Reimagining Futures' was primarily a study of embodiment. Even as the framing research questions focused on literacies, play and community, the attentive eye turned toward the ways that the young people moved through the programme space, interacted with other youth and adults, and marked temporal shifts as their bodies cast off the school day's tensions, transitioning into the afterschool space of Voices (see Appendix for the list of projects).

An attention to embodiment involves noticing children and young people's relationships with the material conditions of their surroundings – playgrounds, classrooms, homes – and how these are borne out in their bodies, and in the ways they carry themselves. For example, in the 'Voices' project, we observed the ability of small gestures to invoke a wholly new context – a circle of play formed in the space opened up by laughter, for instance, and likewise, shoulders slumped against the backs of chairs and hands flying across smartphone touchscreens searching for a song to shift the mood of an activity. These embodied actions in the material world tell us something about the children and young people who occupied these classrooms, playgrounds, streets, libraries, and how their lives were entwined with those spaces. Metal and wood, synthetic composites, hardened plastic and sturdy cushioned foam form structures from which children hang, onto which they climb and through which they scurry, tumble and navigate (Kraftl, 2020). Playgrounds, materially and socially, are open to regular transformations in use and purpose such that 'traces of making in the playground are likely to be embodied, ephemeral and fleeting' (Potter and Cowan, 2020, p 251). They 'can be seen as a space where creative, collaborative making occurs ceaselessly in a range of modes' (Potter and Cowan, 2020, p 251). To pay attention to embodiment is to inhabit the unfolding present in a way that is not entirely anticipated by traditional tropes of methodology. This is a poetics of research that is both anticipatory and felt within the moment.

Polyphony

All of the projects discussed here involve a wide range of people: academics from various disciplines, artists working in various media, teachers, social workers, youth workers, children and young people, and members of other groups brought together by shared passions and commitments. People from all these groups write and speak in very different ways. They employ different registers, genres and styles as well as more or less regionally specific forms and structures. In this sense, the projects were polyphonic – a wide range of voices spoke within and through them. This has important implications for a poetics of research. Within the academic context, the journal article is a highly valorized genre, one that typically uses a particular form of language. Its status is connected with the demands of the Research Excellence Framework (in the UK). However, the journal article is of no significance at all in many other contexts. Texts of this kind are of little interest to many artists, many professionals who work with communities, and even less to the children and young people who participate. Hence it is not unusual to hear academics talking about the need to 'translate' their work so that it speaks to other publics.

However, this is to assume:

1. that the research is necessarily to be located in the academic sphere and not elsewhere and hence what happens in community settings isn't really research;
2. that the most important voices within collaborative projects will therefore be academic ones with other voices mediated by those of academics;
3. that a key part of participatory research is translation on the part of academics – academics reframe their research for other audiences but also translate or represent what participants say so that it can be consumed by other academics;
4. that artists are present in large part to create outputs that will be more palatable for 'wider audiences' – there is often little recognition of the artist as an independent thinker.[2]

All of these assumptions need challenging.

1. As the earlier section on 'Work' emphasizes, the activities of the project may take place largely outside the university and result in artefacts such as films (Pahl and Pool, 2021), exhibitions of material objects, small stories[3] or personal testimony[4] – productions that are characteristic of other spaces,

---

[2]   This is a key theme of the 'Co-producing Legacy' project; see Pahl et al describe the project in Facer and Pahl (2017).

[3]   See Pahl and Pool (2021) and the story of a locust on a teacher's back.

[4]   See also Lassiter et al (2004), where African American community members argued that they were not represented in the previous work by anthropologists, called *Middletown: A Study in Modern American Culture* (1929). 'The Other Side of Middletown' (Lassiter et al,

contexts and communities. The challenge for academics is to listen to the voices that speak through these productions and engage with them on their own terms without viewing them as local or parochial (Escott and Pahl, 2019).

2. It may well be that the most important voices are those of children and young people or community members and the way in which those voices speak is not incidental or a consequence of an inability to speak with an academic voice. The modes of speech (in a broad semiotic) sense that are employed in these works are an essential part of what they are and what they do.[5]

3. In this context, the concept of translation – a back-and-forth relaying between academic and non-academic ways of speaking – is not entirely appropriate.

4. Artists also speak 'other' languages – visual languages, poetic languages, languages of action and practice. Their involvement is not always (or indeed even often) about communicating academic ideas palatably. Indeed, their work may be at odds with academic ways of knowing and speaking.[6]

Thus multiple voices speak using different languages (sometimes in the literal sense that participants are multilingual, but also in the sense that they are fluent in other styles, registers, specialist lexicons, local varieties, genres, and vocabularies of imagery and visuality).[7]

In conceptualizing this polyphony, the work of Mikhail Bakhtin is useful. Much of Bakhtin's writing on polyphony is concerned with distinguishing the novel from other genres such as epic, tragedy, or lyric. The distinction lies in how these genres respond to a world in which multiple languages interact, compete, and – as he puts it – illuminate one another. The older, 'straightforward' genres, he argues, adopt a singularity of language, whereas the novel incorporates multiple languages and brings them into interaction.

What is important here is that in the novel, unlike the older, 'straightforward' genres, language is not simply a means of representation.

2004) was written in response to this by the community themselves. This influenced how the 'Imagine' team wrote the Rotherham book, which included images and poetry by community members. It was called: *Re-imagining Contested Communities: Connecting Rotherham through Research* (Campbell et al, 2018).

[5] See Lalitha Vasudevan's book chapter with colleagues, 'Participatory, multimodal ethnography with youth' (Rodriguez Kerr et al, 2020), which troubles the concept of documenting a multimodal ethnography. The difficulty of operating with a bounded notion of what research actually is.

[6] Co-producing Legacy Project, Felix Guattari (2000). Kate and Steve used this to think about the role of the artist-in-residence (Pahl and Pool, 2018).

[7] Odd stories that didn't fit with dominant narratives. See Pahl and Pool (2021).

It is an object of representation. The novel shows us languages and invites us to contemplate them. It ironizes them, parodies them and offers them as objects of sentiment, aspiration and ridicule. For Bakhtin (1981), this is connected with the fact that the novel engages with the present (rather than the heroic past) and with the everyday (rather than the elevated and set apart). In Bakhtin's (1981) essay, 'Epic and novel', he writes that 'The novel … is associated with the eternally living element of unofficial language and unofficial thought (holiday forms, familiar speech, profanation)' (p 20). In Bakhtin's theorization of the novel, laughter plays an important role. He describes laughter as a 'permanent corrective' or a 'critique on the one-sided seriousness of the lofty direct word' (1981, p 55). In this sense it is a 'corrective of reality' in that it is 'always richer, more fundamental and most importantly, *too contradictory and heteroglot* to fit into a high and straightforward genre' (1981, p 55; original emphasis). Thus, laughter is not an incidental by-product of working with communities – it is an integral part of what the work does in relation to reality (see, for example, Escott and Pahl, 2017). The film *Ninja Story*, made by a group of young people (described in Escott and Pahl, 2017), exemplifies this concept. The film mashed together news discourses and action-movie-speak and gesture together with sounds from a number of contexts to describe a world without language. The key point here is that it is precisely in its heteroglot nature that its meaning lies.

We would like to see the projects discussed here as analogous to novels in the sense that they amplify or offer up a range of very different voices speaking in a range of 'languages'. Of course, it is important to listen to what those voices say. But it is also important to attend to them *as* voices and to think about the languages in which they speak. Polyphony is part of a wider picture in which the voices of academia are questioned and decentralized (Patel, 2015).

Worthiness

Another contribution to our account of the poetics of research is the concept of worthiness, which the *Oxford English Dictionary* (OED) defines as '[t]he quality of deserving to be treated in a specified manner; the quality of being worthy to do something'. All too often the educational system attributes worthiness to some children and not to others. Children can be viewed in particular ways, and worthiness bestowed on children who 'fit', linguistically and racially (Baker-Bell, 2020). These injustices can be perpetuated, thus producing an effect of meritocracy, but not its reality.

Journalist Gary Younge (2016), in his book *Another Day in the Death of America: A Chronicle of Ten Short Lives*, writes about victims of gun violence. He cautions against 'the elevation and canonization of the "worthy victim"' (p 48). The implication of his argument is that there are some children and young people who may be regarded as not worthy of resources, care, understanding and, more broadly, empathy.[8]

'Empathy isn't just listening', Leslie Jamison (2014) notes, writing about her experiences acting out illnesses for medical students, 'it's asking the questions whose answers need to be listened to' (p 5). She goes on to make a key point about the necessary embrace of unknowing in empathy: 'Empathy means acknowledging a horizon of context that extends perpetually beyond what you can see' (2014, p 5). An embrace of empathy in qualitative research is both desired and fraught terrain, the latter because empathy risks rendering researchers – who are often viewed as or assume the role of purveyors of expertise – similarly vulnerable to the people and communities with whom they collaborate.

But, what produces empathy? While people may have capacity for empathic actions and responses to one another, they may not choose to act on these dispositions. Why? Rather than theorizing about the voluntary or involuntary nature of empathy, we might focus on the role of worthiness in the proffer of empathy. The link between worthiness and empathy is crucial. If empathy is to be given, the quality of worthiness must also be granted.

Herein lies our argument for the recognition of worthiness as a necessary precursor to calls for empathy. We began with a stance that regards communities as inherently worthy of time, attention and inclusion in the production of knowledge about themselves. This stance plays out in a variety of ways: as time spent getting to know members of a team and a recognition of that effort as not *peripheral* to the work, but rather as the work itself.

In educational research and broader social spheres, we have heard calls for empathy, with recent incarnations taking root in the form of virtual reality immersive experiences to bring distant others closer to unfamiliar ideas or perspectives. Worthiness is reflected in the questions researchers ask and in the ways researchers and collaborators orient ourselves towards those questions. In other words, we need to ask ourselves whether our methodologies

---

[8]   'Children ... read conditions of resource-starved schools as evidence that the state and the nation view them as disposable and, simply, *worthless*' (p 11; emphasis in original) write Michelle Fine and colleagues (2010) in an analysis of the inadequate schooling conditions of young people living in urban centres in the US. In the same issue of *Voices in Urban Education*, Glynda Hull and Jessica Zacher (2010) provide a detailed analysis of a young woman's digital movie as the conduit by which to question the valuing of afterschool programmes that are attended by young people growing up in under-resourced communities. In the video, the young woman questions, 'How much is a life worth?' (p 20) and provides an answer later in the video: 'Priceless'. They are a few of the many researchers who have questioned the value placed on the educational well-being and lives of youth and the communities where they live, and in doing so, implicitly raise the question of how worthiness is built into or decidedly excluded from the experiences of people based on geography, identity and access.

reflect an understanding of our community collaborators as being worthy of redirecting our enquiry or of having their power recognized in either shutting down or reimagining the work we seek to undertake? We see the recognition of worthiness as vital to the poetics of co-produced research and without such a commitment, the reshaping of knowledge in directions that benefit communities will be stilted. Thus, methodological worthiness is not an ephemeral or fantastical pursuit, but rather a grounded ethos that can be evident in researchers' practices.

Audiencing

A poetics of research requires an understanding of polyphony and a recognition of the ethos of worthiness. The two come together when one starts to think about the audience that may receive the work. How will the voices of community members be received in academic settings? How will the voices of children and young people be received by adult members of their communities? We all have experience of the misunderstandings and failures of empathy that can arise when one arranges for particular voices to speak in unfamiliar settings. As the polyvocality of the project is revealed an audience may not recognize the worthiness of the speakers. As a result, the effort to prepare an audience – that is, to provide context about the relationships, materials, journeys and wonderings that produced any particular work – is paramount.

★★★

Here, Richard shares a reflection that puts this tension into better perspective: I've certainly had experiences of trying to share material that has arisen from our projects and finding that the audience isn't receptive.

Some years ago Steve Pool and I created an archive of images – photographs of various kinds of 'street writing' that we took in East Herringthorpe when we first began to work together. I'm using the term 'street writing' to cover a wide range of texts that can be found in public spaces – shop signs, council notices, graffiti of various kinds, memorials, grave stones, notes on the doors of private houses to tell delivery companies that parcels should be left with the neighbours, prohibitions against the playing of ball games, warnings that the ice is thin and dangerous. I remember the day when we took the photographs. It was a brutally cold morning in January and it took a good couple of hours to reach the goal we had set ourselves of collecting a hundred pictures. But by the end of the morning, I think we were both quite excited. The photographs documented a very wide range of texts and textual genres. And, what is more, they seemed embedded in the contexts where we found them in very complex and interesting ways. We subsequently used the pictures in work with children from the local school as a way to get them to think about community literacies, a strategy which proved productive.

However, when we came to present those photographs to audiences, we never seemed able to capture what it was that made them so interesting to us, and to the children.

Some months after our walk in East Herringthorpe, I was talking to Steve and I began to describe a feeling that I had recently experienced. Since I had become interested in 'street writing', I had started to develop a sensitivity to the locations where one was likely to come across good examples. I would be walking around an urban area and it would suddenly strike me that I was entering a space where I was likely to find interesting texts. It was a sort of instinct that emerged through frequent observation. And it struck me that I'd had a similar experience before. I travel a lot by rail and in recent years, I'd begun to notice that one can often see rabbits from a slowly moving train. There's a certain kind of terrain where you're likely to see them – there'll be an open grassy area but with overgrown areas close by, the one perhaps interspersed with the other. And sometimes on a train journey I would think to myself, 'That's the kind of place where one sees rabbits', at which point a couple of them would immediately appear as if summoned by my observation. Steve was interested in the idea that one can develop an instinct for where street writing will appear and thrive. He began to use a metaphor of 'foraging' in the sense of gathering food by searching for it (rather than by hunting or cultivation). (As it happens, the OED lists older and more troubling definitions of the verb 'forage', but we'll come back to that later.)

Steve recorded a friend of his talking about foraging for mushrooms. His friend had exactly the kind of instinct we'd been talking about – a sort of unarticulated sense for which varieties could be found in what conditions and, hence, where were the most promising places to look for them. And then Steve made a film in which the texts from East Herringthorpe formed the visual component, while the account of foraging for mushrooms became the soundtrack. I remember we showed the film at an event we held in Leeds towards the end of the project. (I think the idea was to bring together academics and artists to look at what we'd been up to and see if they wanted to apply for some more funding to do similar things.) There was a woman there who was very critical of what we were saying. Steve and I were talking about the work we'd done and she intervened several times to challenge us. Then, when Steve showed the film, she became quite angry. The film clearly annoyed her, and I think this was because she found it exploitative. She felt that the images had been purloined from a relatively deprived area of South Yorkshire and exhibited for the benefit of those who had taken them. What is more, she disliked the soundtrack, which she described as 'some posh bloke talking about mushrooms' and which – again, I'm inferring this – she saw as an alien text imposed upon an under-represented place.

It's at this point that the earlier meanings of 'forage' become significant. The first definition given by the OED is:

1. transitive. To collect forage from; to overrun (a country) for the purpose of obtaining or destroying supplies; to lay under contribution for forage. Also in a wider sense, to plunder, pillage, ravage.

Hence 'foraging' has not always been understood as a benign activity in which one searches for the food that nature supplies. It has also meant the gathering of food from a civilian population by means of violence, either threatened or enacted, a definition that seems eerily akin to the accusations of the angry woman at the meeting in Leeds.

Looking back on it now, I'm not sure what we thought would happen. Maybe it should have been obvious that to 'collect' materials from a place like East Herringthorpe and exhibit them for scrutiny might well be perceived in negative ways. Perhaps we just shouldn't have done it. But I don't think we intended any disrespect, and what led us to work with those texts was our feeling that that particular area of urban space exhibited a richer kind of textuality than some other more prosperous areas of the town, an intuition that was confirmed by the work that the children subsequently did (Pahl, 2014). So I wonder now whether we could have done something else with the materials or, at least, prepared the audience better to receive them.

★★★

Another example of preparing an audience can be found in work that has involved children and young people presenting or showing films about their research. Many times, we (Kate and Steve) presented films that were not really recognized by their adult audience as serious work (Pahl and Pool, 2017). However, an event (a conference on co-production) which laid out the conditions for children and young people to present as the focus of the conference did maintain an audience that listened to children and young people's voices. Hence another aspect of the poetics of collaborative research is the framing of material for the audiences which are to receive it.

Dis/enchantment

The concept of enchantment becomes relevant to a poetics of research because it constitutes a way of redressing the rationalization and bureaucratization of enquiry, education and, indeed, the social world. In the sociology of Max Weber, modernity is marked by increasing rationalization – the belief that the world is, in principle, knowable, measurable and, hence, subject to human control. Governments focus on the use of rational methods, and the sciences and the social sciences have developed appropriately rational tools which are administered by a growing bureaucracy. Some environments, for example schools, can become entangled in these bureaucratic practices (such as the regime of the Office for Standards in Education (OFSTED) with its target-driven practices and its Key Performance Indicators), thus reducing the ability for schools to explore more open outcomes. Rationalization and bureaucratization are key factors in the disenchantment of the world – the loss or the erosion of a magical worldview. There is a creeping need to adhere to these standards and practices, which are inexorable and hard to resist. Within the field of collaborative interdisciplinary research, they can constitute a barrier in promoting new practices of vitality and imaginative thought as well as involvement of stakeholders from outside the university.

However, wherever there is disenchantment, there can also be re-enchantment. The work of Jane Bennett (2001) titled *The Enchantment of Modern Life* reminds us of this. Re-enchantment in a sense wrests experience back from the rationalizing and bureaucratizing forms of domination (see also Kuby and Rowsell, 2021). Re-enchantment is not a conventional research outcome, but nevertheless the concept of enchantment captures something important about what took place within the research process. As part of a poetics of research, it belongs in a non-rational and sometimes invisible realm, not traditionally one of research, but nevertheless important.

Following are a number of examples of where the idea of enchantment has proved relevant for us. These include 'Play and the magic circle', 'Becoming animal' and 'Dreamworlds'.

Play and the magic circle

Here, we draw on the ideas of Johan Huizinga (1938) for whom play involves establishing a magic circle in which the usual laws of society and maybe even nature are suspended and new rules come into play. When you play, you accept the validity of rules that apply within the magic circle. Thus, play takes place in a special space, an enchanted enclosure, and has the property of asking 'what if'. Just for a time the players inhabit a new space and abide by its rules. So play, for Huizinga, is connected with magic. The 'Language as Talisman' project operated as a provocation – what if language was magical? This invitation was extended to young people, youth workers, teachers and members of the research group. Likewise, in both the 'Reimagining Futures' and 'Education In-Between' projects, play held magical significance in contrast to the usual conceptualization of teenagers' play in terms of danger and disruption. Peter Gray (2011) notes the lack of free play in the structured lives of teenagers, stressing the distinction between childhood play, which is left relatively open, and adolescent play, which is often organized into sports or games with predetermined rules.

A circle of play may be drawn spontaneously, by the slightest shift in gesture or language. For example, a video camera cradled in a young man's arm wordlessly signals the onset of improvised story-making to be video recorded. A disposable plate is transformed into a prop made to resemble an alien spaceship. Characters that are remixed from a popular culture canon seamlessly interact with each other as a new story unfolds, and those in the background respond intuitively with requisite sound effects. Laughter permeated every pocket of the afternoon hours. But play is not laughter, alone. Play can be contemplative, curious, unpredictable and dangerous.

Becoming animal

An aim of some of the projects was to develop a kind of wisdom, rather than to collect data. In the project 'Communicating Wisdom' (see Appendix), the elderly anglers talked about the feeling of 'thinking like a fish' and this was part of their practice as anglers. Likewise, in the 'Odd' project, children habitually became animals (for example, cats) in their dreams, and filmed the switched animal/children as part of their commentary on what it was to feel odd. This 'other-worldly' quality was often child-led and involved a reversal of adult–child relations, whereby children, rather than adults, led the way.

Bennett (2001) in *The Enchantment of Modern Life* writes about the wonder which is generated by 'becomings', and the becoming 'other' as a result of engagement in fishing might exemplify this process. Wonder has an ethical dimension according to Bennett. It is not sufficient to have a moral code; one needs a will or disposition to live by that code and wonder, she argues, is generative of that disposition. As she explains:

> Enchantment is not a moral code, but it might spark a bodily will to enact such a code and foster the presumption of generosity toward those who transgress or question it. My wager is that if you engage certain crossings under propitious conditions, you might find that their dynamism revivifies your wonder at life, their morphings inform your reflections upon freedom, their charm energizes your social conscience, and their flexibility stretches your moral sense of the possible. (Bennett, 2001, p 32)

Bennett's argument is that these other-worldly becomings (such as becoming animal) produce an affective state which leads one to behave and think ethically and become attentive to other people and their ways of being, including animals.

Dreamworlds

Day-dreaming may also put one in touch with an enchanted state, an imagined world in which things are different and perhaps better. Writing in the 1920s–1950s, Ernst Bloch developed a rich and complex account of utopian thought in his three-volume work, *The Principle of Hope* (1986). In this context, he argued that day-dreaming was a utopian practice precisely because it offered a sense of how things might be different, and hence foreshadowed future possibilities. Bloch's ideas of 'forward-dreaming' suggested ways of working in which young people were invited to enter a space of hope. The space of the projects encouraged day-dreaming and a reimagining of past, present and future. This could take the form of using artistic methods and film making to rework past stories, or listen to day-dreams. For example, in the 'Odd' project, children talked of the need to day-dream and to listen to the 'deep back of my brain' (quote from a child) as a way of moving away from the spaces of school. Part of day-dreaming is the work of moving to the 'off-task' realms of imagination and play; these can be productive as well as potentially disruptive and our projects celebrated what they could produce.

Conclusion

In this chapter we have explored the concept of 'poetics' – understood as the 'creative principles informing any literary or social or cultural construction'. We have particularly focused on the idea of collaborative research as an artful endeavour. The unfolding element of the experience is something to focus on. A poetic approach is about creating the space for something to happen. We have explored concepts that enabled the poetics of research to emerge. The concepts highlight aspects of research which are connected to emergence and multiplicity and can open up approaches to co-production with children and young people.

The genealogies of these concepts do not necessarily lie within research methods, but include sources such as art school practice. Work, for example, is an art school concept that enables a focus on process not product. Connections between making and doing are remade in a discussion of 'where the work is'. This also enables a valuing of the unexpected outcome and serendipity. Stories can cross continents and be remade both within research and everyday contexts. They are protean forms of thinking and communicating. They can be understood as modes of knowing and thinking and a way in which knowledge can be appreciated, crafted and shared. As we asked, where is the work, we also asked, 'what is the story' as part of the poetics of research. The concept of embodiment entailed a focus on gesture and the body and this, in turn, expanded what we noticed, and how we worked within contexts. Listening became entangled within a gestural practice. The work has to encompass a wide range of communities and a polyphonic approach to the projects foregrounded listening as the focus of the work. Voices are not always heard if they are couched in unfamiliar discourses, but the work of research is to listen as well as to make and share the work. Sometimes learning how to listen can be the work of the project.

The concept of poetics calls into focus the importance of paying attention to representation. In-between moments of laughter, gestures and silence, become part of the picture. The accidental and incidental become important foci of the research. Paying attention is important. The concept of 'worthiness' creates a focus on who is listened to and why. By linking empathy with worthiness an expansion of thinking is created about who matters in research and why. Learning how to listen to children and young people requires being an audience, but sometimes the audience is not prepared for what they might hear.

Working with a poetics of research means being prepared for the unexpected. The concept of enchantment enables surprise, awe and wonder to come back into research. Within the realm of enchantment, play, becoming animal and day-dreaming can then become modes of working and thinking.

By discussing some of the terms and concepts that make up our poetics of collaborative research, we have prepared the ground for the chapters in this book. Our concepts presented here will reappear in the chapters to follow. What these are are propositions, or topics to bear in mind when undertaking a poetics of research.

INTERLUDE 2

Postcards

Richard Steadman-Jones

I

"I once spent three weeks in the flat without going outside." This is Stan describing a personal low. "The furniture was taunting me."

I think of the table in Stan's kitchen, the radio, the mug-tree, the unmatched chairs pushed up against the wall. Surely he hadn't heard them speaking? "Through their familiarity," he says, tuning into my puzzlement, looking at me earnestly. " 'Still here?' they asked, 'Back here with us? Back here with us, big shot?' Every detail of the bloody place burned into my mind. A brand on a cow's arse."

He smiles and I feel anxious. I look at my hands. If I were Stan, I might recoil from the same old sights – the scar on my thumb, the spot of pigment at the base of my index finger, the scratches in the ring that belonged to my grandfather. I sense how, for him, the habitual might have become a source of danger.

II

The man in the cafe on Wilkinson Street had two stacks of postcards, each held together with a rubber band. He asked for tea and I saw that Sylvia knew him, certainly as well as she knew me. His postcards were no novelty to her.

There was evidently a method. One card to be drawn from each stack. The two to be placed side by side on the table and contemplated. The time to be measured in sips of tea – six or eight to each pairing, fewer unacceptable, more to be avoided, a sign perhaps of self-indulgence? An entry to be made in a notebook, clearly written in the centre of the page. The cards to return to their respective decks. The whole to pass through five or six iterations. This is something to try, I thought. What is there to lose?

III

When Stan met Esme, the bonds of familiarity broke – or at least they frayed. She teased him but not with the cruelty that the chairs and the mug-tree had shown. There was a springy quality to their interactions, as when a green branch is pulled out of position and whips back on release, trembling a little until it reaches equilibrium. She did not seem perturbed by his characterizing the fridge as a giant gob with the vulgar habit of opening to reveal what it had eaten. She even appeared to enjoy the image and sometimes used it herself, though she never let it put her off her breakfast. He enjoyed how she remade the world through the subtle play of her perceptions against his own.

IV

Sylvia knew little about the man with the cards, nothing that wasn't obvious. She saw him as 'practical' – surprising, given the esoteric nature of what she confirmed to be his daily ritual. I had interpreted the practice as a species of divination, like casting the runes or reading the hexagrams. I would have been embarrassed to reveal it to the workmen who were using the cafe while they widened the pavements in Garrison Street or to Sylvia's daughter, who routinely expressed her contempt for the customers in a kind of floor-show of hostility that was widely admired in the area. He, though, did not seem at all embarrassed.

I began to collect postcards without knowing what they were for, picking up a few when I saw some and shuffling them into the growing stack, unsure of the correct way to do it.

V

She was angry in a way that Stan had never seen before. His mother had fought against her rage, powerless but ashamed to let it take her. In the midst of the verbal gunfire, she would wail with a grief that arose from nothing so much as the horror of what was happening to her, the loss of what she felt was her 'normal' placidity. But this new kind of anger was clean and icy, hard and luminous. It *was* Esme, not something from outside her. He heard a few home truths that morning and had to admit that, unlike some of the utterances that go by that name, they were, in fact, remarkably truthful. When she had gone, he stood before the mirror and felt himself caught between the sick familiarity of his face and the possibility of some new arrangement. The quiet sounds of the day beginning. She'd be back in the evening and now was the time to put pen to paper.

VI

Pairs arising. Four views of Cromer + Virginia Woolf. Whitby harbour + a naked rugby team. York Minster + the docks at Grimsby. Stanage Edge + a vintage ad for Oxo. Wild deer + Groucho Marx. The Golden Gate Bridge + a Henry Moore. Tower Bridge + the front at Bournemouth. Las Vegas by night + Bristol from the air. Stevenage New Town + Capri in the 50s. The prom at Llandudno + Angkor Wat. A nuclear power station + a recipe for Scouse. A bang + a whimper. A rock + a hard place. A wing + a prayer. A gasp + applause. A belt + braces. A son + a lover. A man + the people. The devil + the deep blue sea.

VII

Stan sat by the window and counted his breaths. I need tricks not methods. I have taken myself too seriously. A little trickiness is long overdue. The workmen were finishing up, loading the temporary signs onto a truck. A trick and a truck. Trick truck. Trick truck. (The sound of a little horse on a stony road.)

"No cards today?" A stranger stood beside him. For a moment Stan was bewildered. No cards?

"Uh, no!" he said at last. "No cards today." The stranger nodded.

"Is that a permanent development?" This irritated Stan a little.

"I don't know," he said. "Maybe." A pause filled with sunlight and the chuntering of a fridge in need of repair.

"I've been trying it myself," said the stranger, dropping into a chair, his voice becoming urgent. "For a month now," he added. "I've filled a notebook with pairs." Stan let some silence into the space between them. "Such as?" he said at last. "The Blackpool lights," said the stranger, "and Simone de Beauvoir."

"Good one," said Stan. It *was* a good one. His interest was stirring. "Go on," he nodded.

"Sunset in Cleethorpes," said the stranger, "and ...", his voice dropping to a whisper, "*an 18th-century Noh mask.*" He seemed appalled by the horror of the coupling. "Beginner's luck," said Stan. Sylvia, back behind the counter, began to butter bread. "But what do I do with them?" asked the stranger. "What are they for? How do I use them?" He seemed desperate, pleading, begging for instruction. "What do *you* do with them?" he asked again, crowding in closer. The terrible noise of the fridge again. "Just what is it that you *do* with them?"

The little horse on the stony road.

A departure from the crushing familiarity of the ambient day.

Stan gave the stranger a long and level look – amazed, amused, incredulous at what he heard. He ran a hand over his head and laid it palm down on the table. He sat back in his chair and furrowed his brow with the genuine consternation of one who meets a visitor from an unknown star. An eon passed.

...

"But don't you know?" he said at last.

VIII

Disclaimer: Stan, Esme and the I of the narrative are composite characters developed through the process of randomly pairing postcards. They are not to be identified with any real persons, living or dead.

3

Worldizing

Kate Pahl and Steve Pool

> Richard: How does it feel to use theory?
> Kate: That's a good question, as it helps me understand that
>   theory is *lived*, not just an apparatus to take off the shelf.
>   In this chapter, we approach theory in terms of how it
>   works within the lived experience of a project. We have
>   thought about the effect of theory as something that goes
>   in and out of the work and we have suggested some terms
>   that make sense of that.

This book draws on thinking developed within projects that were co-produced with community members, including young people, artists and academics (see Appendix). Steve self-identifies as an artist and Kate as an ethnographer. In order to work between our two disciplines, we often feel we have had to sacrifice disciplinary clarity for the potentials that emerge through close collaboration. While we are located in an Education Department (as a broad field involving children and young people in everyday life as well as in schooling), we draw on principles and theories from fine art and philosophy, as well as ethnography, as a way of finding a language to express what we do and how we do it. Our work moves between the more experiential edges of social science through to art-making. This can involve imagining art-making as a knowledge producing practice. Our work is emergent and provisional and flows across the boundary between art practice and research (Loveless, 2019; Pahl and Pool, 2021). This chapter manifests a tension as we work through what it is to think with different theories within engaged research. By 'engaged research', we mean research that explores the idea of 'living knowledge' (Facer and Enright, 2016). This is knowledge that emerges from lived experience within the everyday (see Chapter 2).

Within most areas of social research, when a doctoral student working in the social sciences or a humanities discipline arrives at a university and begins their study, one of the questions the student will get asked is, "What theory are you using?" This can seem hard when the student has wanted, for many years sometimes, to do research that is about trying to understand a particular social problem and plan actions in response to it. This could be something like challenging power structures and inequality in education, creating spaces for children to play or changing how something is taught in school. The student often wonders, "Why is theory important?", or gets stuck into the work without actually considering what theory is, what it does and how it fits within research. Theory can be bandied around, used as both a carrot and a stick. It can be evoked, as something 'outside' the self, an external framework, or something we experience as deeply connected to the self, congruent with ontologies that are closely held within lived experience.

'You are interested in what theory can do in *real world* problems – I'm interested in what theory does with the problem of the *real world*.' (Text message from Steve to Kate, 2 October 2020)

This text message holds our paradox. Steve is often interested in asking the question that comes before the question that is being asked, and Kate is often interested in the questions that emerge from the question that is being asked. We see this as working backwards from an initial question and forwards from the ideas that begin to emerge. We recognize that theory and practice are not distinct things; yet, for the purpose of this writing, we acknowledge that the way theory and practice are encountered within academic research can generate a separation that can lead to problems for both thinking with theory and working on delivering projects with multiple outcomes. Our aim is to wrestle with theory as we bring it to the world of engaged research and to ask critically what theory can do to help us understand, with others, the world outside of the university. This is not to resist difficult theory or question its relevance; rather, we want to work out how it helps thinking.

We acknowledge the problems that arise from using the term 'theory' within this writing. Although at times theory feels like the best word to fit a broad set of ideas and approaches, it also can be a dangerous catch-all phrase that can end up meaning little. We use 'theory' here in a broad sense to cover epistemology, ontology and philosophy. It can be linked to, but is not necessarily the same as, methodology. In our practice we often say 'theory' when we mean 'the stuff that is not part of the project on the ground'. Theory sits somewhere else. For example, this could be thinking through Critical Race Theory in relation to children's texts or aesthetic theory in relation to their creative productions. However, as we explain later, as students and researchers, we are often asked to explain and justify the theoretical position and origins of our work. We are writing to outline a conflicted and emergent relationship to thinking with theory within the felt and material life of our research lives and lived experiences.

Students at universities are often taught to create a research project through developing a set of clear questions or encountering a social context and then applying theory to it. This is the practice that emerges within the university and develops from a long history of scientific enquiry. Methodologically, empirical work is situating the 'work' within the world, and it provides the touchstone for thinking. Our work has emerged from conversations with people, with places and in practice. For example, Steve is a volunteer at the playground where he undertakes his doctoral study. Kate has a background in outreach work and draws on that in her research work. When we work, we engage in the messy reality of the everyday – that is, the 'orts and fragments' (Woolf, 1978/1941, p 140) of our daily being, whereby we are parents, workers and thinkers, moving between the school playgrounds, staff rooms and (before COVID-19) coffee bars where we did our thinking and working. When we create things with children and young people, we draw on identities that are emplaced within long trajectories of our work but also our lives, and the work is carried out in conversation with these identities.

Working across a number of different projects and negotiating many complex sets of relationships led to a realization that in many cases, *the theory emerges from the project*. We suggest that working out from the project space generates different understandings of the world. Some of these understandings are embodied and in place; they stem from knowing that, for example, the ground beneath us is riddled with tunnels in a coal-mining area, and also knowing histories and stories from an area. Our thinking is developed in conversation with these spatial experiences. This way of working links to the fine art concept of practice-as-research (Nelson, 2013) but also stands on its own within our work, on the borderlands of art and ethnography (Schneider and Wright, 2010. We recognize from Joe Ravetz and Amanda Ravetz (2017) the different modes of thinking that art surfaces, and the resulting shift in understandings of this thinking. This thinking can change a narrative of what universities are 'good at' and the idea of expertise. Instead, we would suggest that the *relationships* between universities and communities are what are important and needed (Pahl, 2016).

When we work together, we talk about the '*stuff*' of the projects. This was a way to avoid words that already hold weight and historic connections in relation to theory, such as *objects* or *things* or *assemblages* or *networks*. The word '*stuff*' holds back the weight of theory for us. *Stuff* is a word that works better in conversation than in written text. Used by Daniel Miller (2009), it evokes our daily lives and the objects within the everyday. It is vital to recognize the importance of dialogue, and it is within this genre of communication that we enter into ordering the *stuff* of this chapter.

Our definition of theory here is that it lies across an idea or experience to place it into a more general schema. For example, Pierre Bourdieu's idea of *habitus* (to be described later) captured both intergenerational thinking and being and the living of life. Theory can illuminate, transform and shift how we think within the work that we do. It can enliven us, irritate us and make us so cross that we throw a book across a room. It is there to be thought with and within. We have deliberately kept away from theory when we wanted to keep close to the ground, but we have not ever been distant from it. Our work has been built from where we worked and thinking emerged from those sites.

Our experience of working in contexts such as the ex-coal-mining communities in South Yorkshire, UK led us to seek approaches that work with and pay attention to communities. In a project ('A Reason to Write'; see Appendix) where we worked with a group of children in a school situated in the ex-coal-mining areas of South Yorkshire (Pahl and Pool, 2011), we argued that the young people produced the thinking and ideas, and we reproduced their words and thoughts as much as possible. The artefact of the 'methods kit' including questionnaires, focus groups and interviews crumbles in the face of children's perceptions and ways of knowing, and their way of thinking can't always be captured by linguistically oriented methods (Gallacher and Gallagher, 2008). Some of the ideas that have brought us to where we are now include the work of John Law, who in *After Method* (2004) suggested that methods, and even more their practices, constructed the field. Therefore, we see theory as a thinking tool not as a solid form. In the subsequent section, we consider ways in which theory has arisen from community and everyday contexts.

This respect for communities and their ways of knowing and being has been the subject of a number of discussions. In their joint discussion, Avery Gordon (2014) and A. Sivanandan considered ways in which theory can emerge within projects. They discuss the idea of 'lived theory'. We learned that there has to be an organic relationship between the experience and its meaning for it to lead to action. In other words, there has to be an organic relation between theory and practice – a relationship that takes in the general (state, society, economy and so forth) and the particular (the individual, the community and so forth) both at once, moving between the two levels – seeing the general in the particular and the particular in the general; the wood in the trees and the trees in the wood. For Sivanandan, 'our thinking is flexible – because we are not tied to long-term research projects but essentially addressing ourselves to problems on the ground' (Gordon, 2014, p 5). We found the idea of moving between two levels helpful when we thought about the experience of theory, and the idea of 'lived theory' links to our site-based work.

In our work together, we have been experimenting with what theory brought us. When working together, we have found the idea of 'bricolage'

(Rogers, 2012) helpful. When working in a school or a playground, we often turned to what was to hand; we accessed what we could when we could. This felt like a 'hand-made' version of theory, which surfaced in text messages and in-between conversations in the car on the way to the school. Rogers described bricolage as denoting the 'methodological practices explicitly based on notions of eclecticism, emergent design, flexibility and plurality' (2012, p 1). Intellectual bricoleurs work with what there is, with fluidity and feasibility, drawing on Levi-Strauss' idea of the bricoleur as author of their own curated story. Matt Rogers (2012) suggests the idea of the bricoleur can be useful in moving away from 'rational' ways of doing things, but instead opens out new routes to meaning. Bricolage implies an orientation to experimentation, multiple meanings and a playful account of methodological innovation in an interdisciplinary space. Part of the playfulness of these kinds of accounts of research recognizes the multidimensional nature of research encounters, rooted in mess (Thomas-Hughes, 2018) and structured in a relational, open way. A narrative bricoleur might steer away from uniform representations and, as we described in Chapter 2, adopt a polyphonic perspective.

When Kate was working in Rotherham, she began to understand that the idea of the 'lone ethnographer' did not work in the sense that the meanings being generated in the research were generated within groups of co-researchers who were situated inside and outside the university. She used the idea of 'collaborative ethnography' from Eric Luke Lassiter (2005) to recognize this process. Being flexible with methods and playing with their potential have always been part of our approach. Recognizing where we are and what we are thinking (particularly post-COVID-19), has been a necessary part of the work. Our work has drawn on collaborative ethnography (Campbell and Lassiter, 2014) as a way to bring in a mode of working that engages all participants in the framing of research questions, collecting of empirical information and process of co-analysis. In this way, a work of research becomes something more shared, more contingent on everyday lived experience, and less detached from where it was explored and shared. This work became the book *Re-imagining Contested Communities*, which represented Rotherham through the eyes of the people who lived there and who co-researched with us (Campbell et al, 2018).

We have argued that sites and spaces generate the theory (Pahl and Pool, 2018). Working in sites where expertise is held within contexts such as artistic or digital work requires ways of thinking that locate and recognize that process. Grounded theory is also an approach that recognizes that theory can come after an engagement with a research site (Urquhart, 2012). While theory is emergent in the process, it can be used later on in a research study, while the researcher sits with the data. This approach has a view that data can provide the initial lens from which to think. This sometimes involved a casting off of existing structural ideas, to test the mess, experience the 'strategic scruffiness'

(Phillips and Kara, 2021, p 171) of the process that this involves. Our approach has been to think in spaces and to find ways of thinking within that grounded space-thinking. This is the work that we describe here.

Much of what we did involved co-creating and making with children and young people, with their production processes foregrounded in the projects. As we worked with children and young people, we became concerned to value their processes of making and doing *as* research. We encountered research-creation when working on the 'Odd' project (see Appendix) (Pahl and Pool, 2021). We argued that to value the work as 'the work' we needed to recognize research-creation as a mode of approaching artistically informed productive work. Natalie Loveless (2019) argues for the idea of research-creation as a way of articulating a mode of working that is in tandem with other practice-based approaches in the arts. Research-creation, as configured in the context of this thinking, enabled the shift from particular structures of research and thinking to a more process-led, emergent mode of enquiry, drawing on the work of Erin Manning and Brian Massumi (2014). Research-creation can enable 'experimental and dissonant forms of practice, research and pedagogy' (Loveless, 2019, p 4). It offers a more playful and less stratified model of research practice that works against binary structures and with everyday knowings, stories and embodied ways of knowing. Tangible and intangible forms of knowledge – including material forms, dreaming, playing and making – become enmeshed in the forms of knowledge production. This offers an approach to research 'rooted in process' (Loveless, 2019, p 25). Thinking through the projects has enabled us to articulate our own relationship to theory as a living process, and its emergence within the projects has been something we have traced within these stories.

Forward dreaming: working with theory

We have found it helpful when we write together to work through examples or a telling tale from our work as it can help to create new thoughts and connections. Here is an example of where a site – in this case, a modernist housing project – generated a theoretical stance, which was found in the lived experiences of the residents, but was shared by the university academics:

> An example of why we turn to theory came to light in a university/community project called 'Imagine' (see Appendix) that explored the experiences of the residents of a large-scale modernist development in Sheffield called Park Hill. This was a housing development built in the utopian moment in the late 1950s when social housing was a positive step forward in the UK.
>
> The project team (Kate, Prue Chiles, Louise Ritchie, Paul Allender and David Bell) interviewed the residents and worked with them to understand

the lived experience of Park Hill. Kate with her colleagues (Chiles et al, 2018) thought about the idea of 'forward dreaming' from Ernst Bloch (1986) in order to describe the ways in which the residents altered their spaces in their daily life. From this came an approach that recognized the material and embodied nature of living in Park Hill but also understood the experience of living there as a form of reimagining (Chiles et al, 2018).

A model-making workshop, whereby small-scale models of the flats could be changed and rearranged, enabled the residents to materially reimagine their spaces, and through this the 'forward dreaming' described by Bloch (1986) was actualized. Theory here came into an embodied and material experience of thinking how places could be. This led to the recognition of 'the importance of "daydreaming" when thinking about transforming communities' (Chiles et al, 2018, p 133). Theory here became a way of articulating and pinpointing complex experiential modes of being and thinking. It both was situated in the place of the project (a utopian housing project) and in the thinking of the residents and the conceptual understandings of the academics. It was felt, bodily, and in the 'concrete utopia' of the buildings, through the lived experience of the residents.

This description of a project and its theory shows how a site, a space and an experience can trigger an encounter with theory that is located in feelings. These feelings were connected to the rough concrete surfaces of the flats in Park Hill, recently remodelled and redesigned but retaining their brutalist flavour, but also connected to the wider feelings of the original architects and their visions of a better future through social housing. Theory emerges as useful in this site–specific project.

Two stories and a realization

To try to explain how theory comes into the lives of our projects, we now tell two personal stories about moments of realization where the reading of theory directly connected to how we understand the world of our projects. We choose examples that have had a significant impact on both how we work together on research and how we think about the nature of the world. In the middle of the process of making and doing within the thick of it, new concepts of research and knowing emerged. These ways of knowing and doing might not take the form of 'research' in a university. Ideas and thoughts were fuzzy and diffuse, and at times the collisions of thoughts became inherently messy, crossing borders between the concrete experience in places in communities and the reading we were doing in the university. The 'finding out' of research and its newness often lay in the mix of encounters between the everyday worlds of youth work, art practice, reading and thinking together, and action. The task was not to know something from the outside but to encourage everyone to be part of building knowledge together.

Kate's story

In this story, I (Kate) write about how I have used theory in my life. I am writing this as a bit of a stream-of-consciousness type thing. It is Monday morning, we are threatened not only with a new lockdown but existential angst in the form of Trump getting in again.

When I started my PhD, I found theory hard to access. I had to read books many times before I understood them. One book I liked was Bourdieu's (1977) *Outline of a Theory of Practice*. Bourdieu begins this book by talking about the outsider nature of the anthropologist, who in a way has to provide a map to make sense of what they are experiencing. He talked about the gulf between the potential abstract space and the practical space of journeys actually being made. Bourdieu is concerned that interpretative processes are themselves not the object of study, and he argues that objectivist knowledge, in the form of theory, can distance the viewer from the subjects of research.

In my case, I would say theory arose from the ground of research practice. I spent three years sitting on the floor in front rooms of homes, inhabited by boys who, aged five, had experienced some disjuncture from school. Looking back (this was the late 1990s and early 2000s), I had a utopian view of the boys as misunderstood by school, and when I encountered them (or their parents), it was initially accompanied by an idea that school had somehow had a problem with the boys and my research would fix this. However, my supervisor, Brian Street, asked me to do something different, which was to find out "what is going on here". Accordingly, I collected Pokémon cards, followed treasure maps and created ways in which I could reciprocally understand the households I visited every other week in the early evening. My field visits were often accompanied by food and other chats. I would sit on the settee, learn about what the children had done and soak up ideas from the households. In a naïve way, I thought I was doing fieldwork.

After about two years, I realized that I had begun to know what would happen when I turned up on the doorsteps of the families. I knew about the train built by a grandparent as part of his work building the Indian train network. I knew about the special food produced at Eid al-Fitr and the importance of the display cabinet for the mini 'Fimo' Pokémon models as it had emerged from the field research. I began to locate my thinking in the norms, practices, material objects and stories that themselves made up the homes and the meaning making of the young children. I turned to the idea of 'habitus' from Bourdieu as it made sense of practice; it was grounded in practice, but it was not practice. It provided a space to think – to have a dialectical relation between the 'objective structured dispositions within which those structures are actualized' (Bourdieu, 1977, p 3). Bourdieu's theory is helpful because much of his thinking turns on the relationship the researcher has to the field (in my case, the homes) (Grenfell and Pahl, 2019). One of his key contributions was an interpretative schema that accounts for the process of observing and the living of life. This idea was the habitus.

One of the difficulties I have encountered in academic discourses is that behind many ideas lie genealogies of practice, that themselves are hidden. Bourdieu, in a book with the sociologist Loic Wacquant, described the process of unearthing 'the epistemological unconscious' (Bourdieu and Wacquant, 1992, p 41) as important in order to understand how thinking is developed in the field. The concept of habitus grew out of a long enduring encounter with the Kabyle people in Algeria. Moving in and between fieldnotes and thick description, Bourdieu's concept of habitus was both lived and written.

What I liked about habitus as a theory was its improvisatory quality. It was enacted in the living of the world. Bourdieu (1977) called the habitus 'the durably installed generative principle of regulated improvisations' (p 78) and he understood them as history wrought through everyday life. This created an understanding of how the past worked in the present. In my fieldwork I observed how stories from grandparents came together with current popular cultural obsessions – for example, the figure of Super-Mario was combined with a story of a bird from a memory of chasing birds in the child's grandparents' village in Turkey (Pahl, 2002). Habitus described two things in one – intergenerational knowledge and practice, and the 'now' of making. Habitus was a kind of shorthand to describe what is now and what was then, and the relationship between the two, in a home setting. It also described embodiment – the leap of a wrestling encounter and the arc of a pen, as well as food and ways of doing things over time, every day. It could be scaled up, patterned and turned into a way of understanding intergenerational knowledge in homes. It also described the process of learning in the home. This is an embodied process:

> [T]he 'book' from which the children learn their vision of the world is read with the body, in and through the movements and displacements which made the space within which they are enacted as much as they are made by it. (Bourdieu, 1977, p 90)

Crucially, Bourdieu also described the 'magic of the world of objects' in the same breath – he did not separate out language and material culture.

As someone interested in the potential of cultural materialism and the power of things speaking, I found habitus a wonderful word to describe how 'each practice comes to be invested with an objective meaning' (p 91). It was this meaningfulness that was important to me. Habitus is neither set nor totally improvisatory; it allows for the space in-between for human beings to play.

Here, we pull away from this vignette to reflect on what it means to us. Steve reflects on the work Kate describes from his point of view:

Kate's story intersects with theory on a number of levels. Bourdieu drew on fieldwork to develop a theory of habitus from his extensive contact with the field, and this then led to the idea of habitus as a way of thinking about how people live their lives across generations. Habitus is part of an overarching theory of human practice that changed the way Kate thought through her fieldwork. All encounters with theory do not have to change us as people; yet, some encounters require a shift in who we are and how we understand ourselves. Habitus was a theory that was worked through practice, and it allowed a movement between the on-the-ground study and thinking about it. The relationship between thinking and doing was encapsulated in one idea. We were working with two things at the same time and in conversation with those two things. Through working through the notion of habitus as a dialectical way of thinking, the interrelations within her study changed in nature; they were more than data and more than research encounters. Theory here has the potential to shake things up a bit, to encourage a thinking beyond the space of practice, but not to detract from what is going on here.

Steve's story

I am writing this now on the 4th of November, the morning after the 2020 American election – the result of which is still undecided although President Donald Trump has claimed an unconfirmed and unvalidated victory. The weight of this situation on my world is both heavy and unbearably light. I am reminded of John Berger and his suggestion that at the most difficult times, hope is something we need to bite on. So with hope between my teeth, I will write of John Dewey, Venice and pragmatism.

In the winter of 2019, I was working in a school in Venice on a contemporary art project called 'Neverland'. We were collectively imagining what Europe would look like in 200 years' time. In the evenings I would return to my room and read Dewey's 1935 book on

Art as Experience. I cannot separate my time in Venice from reading Dewey; the two things are a single memory as both of them are under my skin, like an itch, like a wound, like a body. One morning at six, in a rare moment of Venetian quiet before the tourists descend, when walking down the Strado Nova on my way to teach on the mainland, I fell into a rare moment of reverie. In reflecting on where I was and what I was doing and, importantly, thinking with Dewey, I had found a clear insight into the work I do. The words that I can write can describe the idea but only really point towards my feelings at that moment. Dewey's idea of experience became the way that time unfolds; at that moment, any opposition between the product and process within art-making, a subject that haunts my discipline, became redundant. Art is experience. It is immanent and unfolding, never fixed in a person or an object or a moment. Venice, like nowhere else I have visited, has a way of placing you in time, in both its long history, its fragility and its entropy. I had spent years arguing for process, and within the argument establishing the problem; in arguing for process I set it against product and, in so doing, generated the possibility that art could be anything other than process orientated. The exhibition I was working on with young people, the art Biennale happening across the island, everything I was making and had made changed in a subtle yet fundamental way. This was an encounter with theory that changed the way it was possible for me to think about, write about or make art. It felt fundamental and liberatory. Now I carry this feeling about art, time and experience with me and, like a good pragmatist, put it to use in order to help to open up rather than close down how I am able to work with art and people in the world. This moment of realization was a perfect storm, a coming together of work, thoughts, moments, place and time. I felt that I shared in an articulation or an enunciation through Dewey something I was aware of before which had always been opaque, never really spoken, even through an internal voice. I had not known I'd actually known it before I knew it.

I will stop there as I'm not sure if this story is quite right, although I like the fact that it's not directly in a project but it's the slipping out of projects into life – I cannot be more explicit about this as I am already too close to it.

In both these examples, theory slipped into the rhythm and practice of the everyday and made a difference. Working within and outside the world, the theory entered the world with our experience and illuminated it. We used theory as a term to identify thinking that challenges fundamental ontological concerns about the nature of knowledge. We make a space for theory that is not separate and distinct from other aspects of our work but goes through it. In this way, theory does not sit below or above. It neither existed only as text or language, nor is it the culmination of following a trajectory of thought. We do not see ourselves as highly theoretical researchers; yet, we have both come to a point where just *doing* art projects or ethnographic studies on their own terms was not enough for us. We look for something *more*. This *more* in thought is roughly what we are here presenting as theory, although it could have many names. Our attempt to think with theory has led us to some further thoughts about how theory moves in and out of the world.

Worldizing: a manifesto for working with ideas

We propose an idea that can help in thinking about how theory can be understood or encountered in on-the-ground projects. Rather than see this as a technique or way of doing things, we see this as a mechanism to work with two different things that are separate but have to be brought together at one point. We propose the idea (borrowed from the world of sound engineering) of 'worldizing'. 'Worldizing' refers to how sound engineers can create a soundtrack that can move in and out of audio focus – what its inventor Walter Murch called an 'audio depth of field'.[1] The idea that theory could fall into the background only to surge into the foreground when the situation requires helps us to be neither directed by theoretical dogma nor to ignore its significance. Steve writes about it here in a blog post (nd; cited in Pahl and Pool, 2018, p 7):

> The great sound engineer Walter Murch coined the term 'Worldizing' while working with George Lucas on the film *American Graffiti* in 1973. He was struggling to balance the sounds of Wolfman Jack's radio show, playing on young people's car radios across the city, with the film's dialogue. Eventually he took the soundtrack out into the street, played it through a speaker then re-recorded the sound from down the street while randomly moving the microphone. This process blurred the edges of the sound and allowed it to slip into the background; it mimicked the way we hear things in the world.

In the moving or still image, the 'depth of field' is the area of the picture that is in clear focus. For example, in an image with a short depth of field, the tip of a person's nose could be in focus while their ears could be out of focus; an image with a long depth of field could be a landscape where everything from a blade of grass in the foreground to a distant mountain are

---

[1] See https://studyingsound.org/documents/reading/4_Worldizing%20_Take_Studio_Rec ordings.pdf and www.openculture.com/2020/02/how-walter-murch-revolutionized-the-sound-of-modern-cinema.html.

in crisp focus. In films it can be used to draw attention to what is important in terms of narrative, or perhaps points to what may become important in the future. The appeal of the metaphor is not to bring theory into a project as a distinct line of thought or to position theory in conversation with something else. Rather, the idea is to hold theory in the frame of a project and have control over its influence, an ability to generate fine adjustments to the theoretical depth of field. As with film, the aim in the end is to create a single soundtrack that works with the content. It may be a careful mix of different things, but in the end it mixed into a single track containing many elements that hold together.

An example of 'worldizing'

We now follow 'worldizing' into an example. We see this example as one that can show the potential of living within theory and moving in and out of theory. Our project 'Communicating Wisdom' was concerned with young people's mental health (Pahl et al, 2017) but also led to a number of blogs, films and the hypertext (see Chapter 7). Here, we relate the ideas in this chapter to this project (see Appendix).

Steve's story

When we began the 'Communicating Wisdom: Fishing and Youth Work' project, we brought together a number of different threads that included place, people and ideas. Using Izaak Walton and Charles Cotton's book (written in 1653) entitled *The Compleat Angler* as a starting point, we established a team that included young people, youth workers, an artist, a poet, a philosopher, an historian of ideas, a literacy scholar, an ethnographer and a group of fishermen. This group met over five months on the side of fishing ponds in Rotherham.

To expand on our definition of theory and what it does in projects and, in particular, to illustrate the concept of worldizing, we started with the writing of Bloch and his ideas of wisdom. Bloch is a philosopher known for his works on utopia – in particular, the three-volume work *The Principle of Hope*. When we began, we had no plan for how Bloch's ideas would flow into the project and emerge while fishing by the pond. Looking through a retrospective lens in order to expand on how worldizing was put to work is not simply a case of trying to spot or fish for theory. To think through worldizing is to listen to the soundtrack of the project and identify some of the places where wisdom and Bloch's ideas of hope come into audio focus.

To sit and fish, to bait the hook and watch the float bob up and down requires elements of hope and wisdom. Together, these add up to the belief that you will catch a fish, partly as you know from the wisdom of experience that you have caught fish in a similar way and partly from the hope that today will be the day you haul a big catch. Fishing is a pastime more than many activities. It functions to remove us from what we are doing and place us somewhere else.

Fishing is an absorbing pastime. There is value in the fact that the wisdom of fishing originates in the necessity to exclude everything else. A line from a film where a young person says "it's just fishing isn't it" captures the feel for the activity; the fact it doesn't have to be anything else is the appeal and the inherent wisdom of a day spent fishing.

In the time spent sitting by the pond listening to the birds and watching a grass snake swim by, the noise of theory fell into the background. We all listened to the stories of our mentors, the stories of fishing trips and ground bait, of tying hooks and the persistence and focus that eventually pays off. Yet, theory was still present; we were mostly just fishing, but all of us were also thinking about the ebbs and flows of our lives. I remembered fishing as a child and the sensation of an eel wrapping itself around my arm. I filmed a fish gasping for water and wondered about death and responsibility. Richard would write a blog post every morning. I remember the story of a fisherman who did not use hook or bait but sat by the river dangling his line waiting for wisdom. This is now part of a hypertext in my mind; it jumps into crisp focus when it needs to, but does not act as a distraction in opposition to the rest of what's going on.

In a session at a youth centre we all come together to prepare tackle. I film one of the coach's hands as he shows one of the young people how to tie a hook. The two sit in silence carefully mirroring each other's hand movements. I think that wisdom is something that lives in bodies and time; it is not about knowledge or intelligence or anything that you can tell anyone else.

In a room at the university, Johan Siebers, a philosopher, tells us about what Bloch said about wisdom. It starts with the birth of Western philosophy in Ancient Greece. The only thing to oppose philosophy or to deal with the problems of philosophy is philosophy; it generates the problem and provides the only way to tackle itself. This makes sense in a way; I decide to skip the next day and go fishing but theory has come into close focus and I go to the second day and struggle to make the ideas fit. I am fishing with an un-baited line without a hook and perhaps I have found wisdom without expecting to catch it.

This example revealed the way in which an activity – fishing – and a seminar in a university, began to make sense across those separate spaces. 'Worldizing' was an attempt to describe a way of doing things that allowed theory to come in as part of the mix. It also neatly side-stepped the problem we faced in some of our projects, where the concepts of discipline and field were not relevant. Theory and genealogy were less important than the topic in hand, whether it was the nature of wisdom or the need to defend everyday language. The idea of 'worldizing' enabled us to break disciplinary boundaries and focus on the *experience* of theory, not the use of it in abstract ways. It was also a useful word to describe the process of learning together. Theory could be applied where it felt relevant, but not where it was not relevant.

While it is not a problem to bring theory into communities, it is a problem when dense theory is brought into situations where it needs to go away. The idea of 'worldizing' allowed this to happen. Theory can become like an ambient sound that comes in and out of the project ideas as and when it is useful. Rather than a gloss to apply at the end of the project, it becomes a diffusion that can help augment or intensify current thinking. 'Worldizing' enables us to keep theory at bay when needed. That's the worldizing concept. It enables us to break the flow of one mode of thinking so it doesn't suck everything along in that direction – to open a space for the world to break back in. We have tried to provide the idea of 'worldizing' in order to help think through the dominance of theory. We are trying to describe the falling out of things, and the ways in which it tries to simplify things. Perhaps this doesn't clarify things, but it does make us question how theory and practice intersect.

What, then, is the role of theory in collaborative interdisciplinary projects? For us, theories are things – alive, moving, active and fecund with potential, they are things on, of and in the world, things to think with, things to critique, things to challenge, drivers of change and the gateways to dead ends. Our projects had a number of dimensions, but they were all rooted in communities and all were interdisciplinary. We encountered theory in a personal way. Working through ideas together helped us expand our understandings of what theory could do for us.

What happens when a project moves seamlessly between words written on a page to a moment when you are sitting by a riverbank carefully watching the movement of a float, anticipating a bite? Our feeling is that it is an unfolding process. We worked with a sense of unfolding experiences. Part of the reason for coming up with the idea of 'worldizing' was that when we look back on the projects and the forms that emerged, we found it hard to characterize or describe the processes, actions and things that came together to make something new and different. We found existing ideas and modes of description would not hold the forms or the work of this knowledge production.

People know things; they are the experts in the work and the form of their own lives, and are full of rich experience encoded within the mesh of life, the twine of a fishing rod or the child playing with a cardboard box. The work and the form of these projects are alive and always in process. They are not categories of knowledge or approaches; one does not lead to or contain the other. They offer ways of thinking that bring with them different problems and edges to contend with. What we both acknowledge is that there needs to be a different language of description for the process of producing theory within and using theory alongside collaborative work with communities. Our work here is sustained by conversations, by reading, and by thinking. Theory is part of this, but its work in the projects is sometimes visible, and other times less so.

A conversation

Steve: (in conversation with Kate) ... so the thing about worldizing that I think I wanted to say is, I don't think it was ever something that we were supposed to notice happening. It wasn't something we were supposed to do. Does that make sense? I'm not going to call it a method or anything. So it wasn't like looking back on the projects and saying that was a bit like worldizing. It came more from the problem of looking back on the projects and thinking how do we work with some theory in this really practice-rich space within these projects?

Kate: I think that's one of the problems – our projects were quite unusually messy. I think a lot of people just thought that they're a bit fuzzy. But I think that's what I'm trying to identify. That's why worldizing is so important. Well, it's really useful.

Steve: The line between delivering a project and writing it up is very messy – we're not just saying after Law (2004) that the world is messy. It's like the actual messiness of it all seems to be really necessary for us.

Conclusion

In this chapter we thought about how theory is used within lived experience. We explored the relationship between theory and practice in on-the-ground projects, and focused on how theory can become useful within projects. Working through theory can be difficult and we came up with an idea of theory as emergent within the projects. This enabled us to situate theory differently. We focused on a relational approach to theory-making within community contexts. Theory can lift experience into something that can be described across contexts. When methods are an artefact of sociocultural processes, theory becomes a way of understanding the field in a new way. Theory can be held and explored by multiple researchers, with a built-in co-production process enabling theory to link strongly to the field (Campbell et al, 2018). We have proposed a playfulness around theory that can engender new directions and thinking within the work. Research-creation can open up a space for thinking that develops thinking and doing in a space (Loveless, 2019). Theory can emerge from a site, and its contours can aesthetically inform research-creation. The feel of concrete or the sensory experience of a tree can develop theoretical insights.

How can theory be used in research?

When reading this chapter, you might like to think about what theory you have found useful. Much as Kate described an encounter with theory, how have you, as a reader, encountered theory. What has propelled you on in your thinking? When has theory become important? How does it move and stir you to do research differently? If you think about some fieldwork or an experience you have had, what are the key things that spring to your mind – the practices, the power struggles or the ways of knowing? Theory can help with these things. Theory can also move you on – as the improvisatory nature of habitus worked for Kate. The oscillating nature of this process can be fruitful as research ideas emerge. Research thinking can be embodied – it can involve epiphanies where we stop and think and then share the ideas as moving through our minds and bodies. Thinking through theory might mean:

- stopping and thinking about moments when we change our mind;
- reflecting on our journeys with theory using fieldnotes;
- making connections between sites and spaces;
- learning how to use the language of theory within research.

When we add the idea of worldizing into this mix, working with theory can be recognized as a process of things going in and out of focus. As with the fishing example in this chapter, the work is often located in a moment within a field, within a moment, within emergence. Our work here has a provisional quality and this sense of vulnerability is important when thinking about what theory can help us with.

Letting Go

Hugh Escott

What clings to us

This book is about 'letting go' of disciplinary identities and hang-ups in the midst of collaboration. There is a sense here that there are already some considerable things that academic[1] researchers have to take on, or find a way to access, before they can then go about the work of 'letting-go'. Yet, they must then do this 'letting-go' in order to be able to engage with how communities live and understand the world. In the midst of this, I think it would be useful to turn this idea of 'letting-go' around and consider the things that we potentially have to allow to 'cling to us' to be able to participate in academic research. My perspective emerges from my own experiences of working on interdisciplinary and collaborative projects as a research assistant (a role that can often involve holding things together), while also completing my PhD studies.

We can consider 'research' generally as 'a purposeful and systematic investigation that seeks to build new knowledge' (Campbell, 2018, p 87). While research is about investigation, the institutionally situated nature of academic research means that the apparatus involved in investigation involves academic discourses, institutional ideologies, concerns with prestige and beliefs about value. This means that 'new knowledge' emerges from 'artificial' processes. Essentially what I would like to propose is that we consider research as 'artifice'; in the sense that it involves craftsmanship, cunning and construction, is fundamentally contrived through academic discourses and

---

[1] I use the term 'academic' in a general sense to refer to those who are employed, predominantly, by higher education research institutions and are involved in teaching and research.

is institutionally situated. My main points and arguments are not necessarily new (see, for example, Street, 1995; discussed in Chapter 1). However, in thinking about how collaboration challenges disciplinary identities, I simply wish to bring into focus aspects that influence the practices and values of academic research.

I wish to emphasize how 'institutionally situated' research knowledge is produced through particular processes and practices. It is important to try and delineate some of what this entails so that we can not only consider interdisciplinary collaborative research as 'different' (of equal value) not 'other' (of less value), but also, we can consider that 'doing research' is intrinsically tied up with having an 'academic career' because of the institutionally situated nature of 'academic research'. 'Letting go' is potentially needed because the process of collaboration can lead to a recognition that the academic institutionally situated ways of doing things that we value may not be up to the task of engaging with the reality of the contexts in which our collaborators are situated. So this may cause a recalibration of the tools, processes and negotiations that we have learned are central to working in the institutional contexts of academic research.

In the following section, I describe an event that occurred on a research project, and then reflect on the values and processes involved in 'working on' this event to turn it into a piece of potential academic research writing. In doing so I consider:

- craftsmanship: what tools I had/have to attend to this event;
- construction: that writing academic texts in relation to this event involves a contrived process of construction;
- cunning: that linearity and coherence in academic research are tied up with the skillful negotiations that serve to support academic careers.

The 'Shandy Bass Incident'

Back in 2013, I was working on a project called 'Communicating Wisdom' (see Appendix). This was a project looking at the ways in which youth workers and angling coaches created contemplative, hopeful spaces for young people through fishing. These spaces seemed to provide the context in which the young people and adults could engage in the intergenerational communication of wisdom. On a practical level, the youth workers took the children angling as a way of providing them with what they felt was a positive environment (in which the young people could be engaged in important conversations by the workers if they felt they wanted to), and that kept them away from less positive activities. On this project, we were struck by the ways in which children, angling coaches and youth workers would all try to articulate something striking about engaging in complex and equipment-focused actions in order to hopefully achieve the goal of catching a fish. Regardless of whether they did or didn't catch a fish, they would just carry on fishing. The activity of fishing was almost a vehicle for something else. The purpose of the whole activity was actually just to be doing it, 'passing the time' or 'being in the world'. However, having said all of this, while some participants talked about how 'fishing calmed them down', it was also described by young people and anglers as 'just fishing', and that it is not about something else happening. The collaborative outputs of this project involved a number of films created by or with the young people in collaboration with Steve Pool and myself.

I would go coarse fishing with angling coaches, young people, youth workers and sometimes Kate or Steve. I had some familiarity with the youth work and Rotherham, UK contexts, but much of what I was seeing was very new to me as someone who wasn't from, and didn't work regularly, in the community context we were in. It was one of the first times that I was involved in ethnographic research, let alone collaborative research, and so I was learning how to think about the 'field' (the cultural context in which you are participating and studying) and 'fieldnotes' (the things you write to record your experiences and which form part of ethnographic data). I was also learning about literacy studies (see Escott and Pahl, 2017), while involved in reading the utopian philosophy of Ernst Bloch (1986) and Izaak Walton and Charles Cotton's *The Compleat Angler* with the rest of the project team. On top of all this, I was a language and literature PhD student trying to finish my thesis on representations of Yorkshire dialect in the works of authors from coal-mining backgrounds, and learning about animal studies and affect from my peers in my study-space. I was particularly troubled at the time by the tension between the inherent violence involved in the act of fishing, and the culture and camaraderie that surrounded these activities. This was the mixture of theories and concepts which served as the backdrop to an event that took place when Steve [Pool] and I were on our way back from fishing and which provoked a discussion between the two of us. What follows is a vignette based on my fieldnotes:

After we got back to Thrybergh [village near the pond] Steve and I drove out of the car park, Steve mentioned 'Bass lager shandies'[2] and we decided that we needed to get a soft drink as things were very hot. I was waiting to turn the car into a service station by Thrybergh, indicating my turn, and Steve commented that he could almost taste the shandy when there was a screech and I looked in the rearview mirror to see a motorcyclist crash into the back of my car. We were

---

[2]   A low-alcohol (0.5 per cent) soft drink brand made with Bass beer and lemonade.

shocked. I got out and my wing mirror fell off. The biker got up and said he was fine. He was on a sort of dirt-bike, and I don't think he was wearing a helmet. We were relieved to see that he was ok, he apologised, we checked again to see whether he was ok, he shook my hand, and we again asked if he was ok. There was a tiny scratch on my car and his bike seemed fine so he drove off.

In the service station and then in the car, Steve and I discussed what we had seen and learned and Steve began talking about the fishing project as a metaphor for life that isn't a metaphor. Steve said that you go through all these practices to catch a fish and then you do and then you carry on and you don't realise that the purpose of the thing is to just do it. I said I was annoyed because this would have to go in my fieldnotes as I was still 'on site' and it was very complicated and we needed to remember it. (Escott, fieldnotes, 7 September 2013)

Kate, Steve and I like to call this the 'Shandy Bass Incident' as if it was a turning point in our lives. I unpack its relevance here. It is an event that happened during a research project when I was employed as a research assistant. As part of a larger academic text, I have presented it here to you as an edited extract from my fieldnotes. Producing fieldnotes and writing a short 'interlude' for an academic text involves engaging with heritages of ethnography and academic writing. Turning this event into something potentially meaningful for others in a research context involves transforming something ephemeral into a static coherent text. This transformation also involves negotiating what is seen as 'significant' by others, and is tied up with the 'ends' that I want to achieve as someone hoping to have a career in academia. Here, then, I want to position the following in ways that are of personal benefit, as part of the 'artifice' of academic discourses:

- transforming;
- understanding valued heritages of discourses; and
- negotiating significance.

Academic discourses, writing and identities

In thinking about this vignette and interdisciplinarity/collaborative enquiry, I find it useful to consider that academic knowledge is contrived in institutional contexts through academic discourses. Academic research outputs traditionally involve written texts (in the form of academic articles, research monographs and edited collections) that present particular narratives about research processes and the world after the lived event – as opposed to the living of everyday life being emergent – and understood simultaneously through multiple epistemologies, ethical framings, subjectivities and individual felt states. To be able to participate in academic research disciplines fundamentally involves participating in academic discourses. Academic research transforms reality and understanding into linear coherent texts (Blommaert and Jie, 2020, p 11), and values the processes and practices that support the creation of these coherent narratives. As Ken Hyland (2009) discusses in the subsequent block quote, these discourses are involved in saying something about, or changing, the world, but also in many issues and agendas tied up with the institutional nature of academia:

> Academic discourses, however, not only work to construct knowledge within academic communities, but to sustain the prestige of these communities with outsiders. On one hand, such discourses carry enormous cultural authority in the wider society about what the natural and human existence are really like: they answer our questions about the world, explain its intricacies, satisfy our curiosities, and improve our futures. They are the guarantors of reliable knowledge, and we place our trust in their unbiased and uncorrupted representations of reality and our faith in their practical effects. On the other hand, these discourses also represent a constant quest for disciplinary status and prestige. Academic disciplines are not uniform or stable but sites of competing individuals, theories and methodologies as alternative perspectives slug it out for recognition and ascendancy. The prestige of a field, and perhaps its independent existence, is often contingent on persuading powerful bodies in the non-academic sphere to provide recognition and resources. Academic discourses are central to this endless struggle to attract more students, more research funding, and more institutional respect within a context of ever-shifting fortunes. (Hyland, 2009, p 14)

Academic institutionally situated research involves creating coherent linear texts, and the value placed on these texts is central to recognition, ascendancy and having an academic career. The development of academic knowledge has arguably as much to do with academics accruing and sustaining prestige, institutional and disciplinary competition, and individual agendas as with providing insight into the world around us. So 'doing research' is tied up with 'being an academic' and this involves doing the things that are privileged by these institutional, academic and disciplinary discourses. Academic discourses and research emerge from the social, cultural and historical contexts of the time, and are tied up with institutional trends, politics, economies and agendas. Cultural politics influence academic discourses and identities, which in turn influence processes of collaboration. There is a tension then in what research is held up as doing (in other words, creating 'new' knowledge, changing the world and so forth) and the reality of the work involved in creating linear coherent narratives, that can be recognized by research and institutional communities. Kate argues elsewhere, in relation to youth civic engagement, that '[p]art of the challenge is that our research mechanisms cannot easily account for the layers of experience that make up young people's civic engagement practices. We are unequal to the task' (Pahl, 2019, p 34). If we consider that the main work of the academic is to produce linear, coherent narratives in the service of accruing prestige, building careers and shoring up the institution, how can we be equal to the task of responding to the needs of communities, and be adequately involved in bringing about change?

In considering this, we can return to the 'Shandy Bass Incident' and make visible various elements relating to tools, transformation and negotiation that are often not made visible in the process of creating linear coherent research narratives.

Craftmanship

In responding to this event, Steve, an artist, started to see going fishing not as a metaphor for something, but just as an authentic way of being in the world for a time. He also felt that as a group, doing research, we were collectively layering and folding stories together to tell, in the words of Emily Dickinson, a 'slant truth' (Dickinson, 1999) about what we were seeing on the riverbank (as opposed to 'capturing' it or 'getting at the heart of it' through research). Richard, a historian of ideas, thought about indirectness in Dickinson's poetry, whereas Johan, a philosopher, described this moment itself as an 'Augenblick', a definitive moment 'in which something emerges and then disappears' (personal communication, July 2013). Drawing on the 'craft' of ethnography, I can present to you an event that occurred, while recognizing that this narrative can only partially engage with the complexity of the cultural context in which it occurred. Each one of us has drawn on the different 'crafts' that we have been trained in, or are disciples of, to respond to this event. These crafts influence our interpretations but also our identities and the nature of our conversations with each other. The incident was of particular significance to us because it seemed to illustrate our shared understanding of Bloch's (1986) concept of the 'darkness of the lived moment' (p 32). The present is impossible to grasp. We are constantly involved in making sense of the world through both our hopeful ideas about what is yet to happen and the fact that our distance from the immediate past allows us to more easily grasp it (and, therefore, create narratives that help us understand what has happened). Talking about Steve and my shock and confusion, the hot day, being outside a petrol station in the UK, the wing mirror falling off, and interacting with the biker is simpler now that time has passed, but it is also only ever a partial retelling of the 'knottedness' of the emergent present. The discussions that this incident provoked serve as an illustration of the various crafts, disciplines and tools that we choose to bring to the interpretation of what we felt was an event that had something to do with the intangibility of the present. Also, as a group, we have access to these means of interpretation because we have had privileged access to spaces in which we can learn the 'craft' of the artist, historian of ideas, philosopher, ethnographer and so forth. Disciplinary areas, and methodologies, are influenced by social, political and cultural aspects. Our access to them, training in them, the sense of identity and affiliation that they give us, and the cultural authority of these heritages play a significant role in how we can mobilize various ways of thinking to build new knowledge or answer questions about the world.

Here, then, ethnographic practices and philosophical theory are part of the 'craft' available to me to say something about this event, and I have these tools personally because of my access to the professional, institutionally situated roles of PhD student and research assistant. The 'knottedness' of the lived moment that we felt was glimpsed in the bike crash resonates with viewing the present as a contextual 'kaleidoscopic, complex, complicated' (Blommaert and Jie, 2020, p 25) patchwork in ethnography, but also with ways of considering the 'situatedness' of social practices from literacy studies, and the importance of process and emergence in arts-practices. In this example, there are diverse heritages of skills and tools that can be mobilized to respond to the contextual emergence and 'precarity' (Stewart, 2012) of cultural life, but it is also important to think about the biographies, access and identities that influence how and why particular heritages are actually mobilized.

Construction

In the previous extract from my fieldnotes, I discuss the fact that, on a scorching day in June, Steve and I were essentially involved in a minor car-accident while Steve dreamed of drinking a refreshing soft-drink/soda. This is not necessarily the kind of 'data' that I was expecting to collect. Also, we had been involved in angling, an activity that can be understood simultaneously as constructive for young people and, yet, predicated on violence towards animals, and now we were driving home. There is nothing in here about what happens on the riverbank among young people, youth workers and angling coaches. At the time, I was learning about how to turn what happens in 'the field' into 'data', and this incident challenged my understanding of the boundary of the 'field' and the work involved in fieldnote writing. It was taking place 'after' being in the 'field' but before I had got home to write up my experiences in my notes. I wasn't planning on 'capturing' it, and it wasn't the kind of data we were 'looking for' – although finding the unexpected or 'failing' to find what you initially expected are important parts of research processes. The crash provoked a conversation that was very significant for Steve and myself (and also became significant for the rest of the group through our subsequent conversations), but what it was about this discussion and its repercussions is not wholly captured in these notes. The crash also provoked the need to turn my experiences into fieldnotes as it expanded what I had been considering to be the 'field' in which I was collecting data. At that moment I was 'out' of the field, hot and tired, and so the thought of trying to work what had happened, and being discussed, into fieldnotes didn't fill me with enthusiasm. Whereas, this would be different if I was just going to go home and tell my partner or friends about the crash. Collecting 'data', exploring materials, interpreting, improvising, developing arguments and considering various perspectives are all parts of the research process, which look different when built into finished linear coherent narratives about research. And often, stuff that happens is made into 'research' through writing.

I am giving you all these various bits of what might seem like extraneous information to highlight that the 'Shandy Bass Incident' has to be 'worked on' for it to be of significance, and this process of transformation is shot through with numerous other elements. Something like ethnographic research involves turning things that happen into data, through writing practices. Also, learning how to negotiate this process of transformation, and making this competence visible to others, is part of the process whereby individuals become credible in the eyes of the academic community and are granted permission to participate in institutionally situated academic research (for example, a PhD student gaining their doctorate).

Cunning

On reflection, this was a significant event for us at the time, and even now. But in another context, you might think soft-drinks and minor accidents are not significant. I can see the headline now: 'University researcher spends taxpayers' money fishing, buying soft drinks, and knocking people off their bikes'. This event didn't feature in the outputs of the project that satisfied the funding body's criteria and wasn't the focus of our main presentations about this project. And yet, it was a point of crystallization for our thinking, and led to a positive development in our group dynamic and work. Ethnographic research involves constructing narratives, but even more generally, the traditional humanities or social science essays involve a particular narrative structure (for example, introduction, literature review, methodology and so forth). Stories don't have much significance if they don't speak to the audiences that hear them.

In a similar sense to how Pierre Bourdieu (2010: xxvi) argued that it is the audience's eye that creates the value of art, it is how well something can be seen as significant/meaningful to the 'powerful bodies in the non-academic sphere' (Hyland, 2009, p 14) and to those with disciplinary and institutional prestige, that creates the value of research. With the political, cultural and economic landscape of these academic and non-academic forces subject to constant shifts over time, creating different norms and values have to be negotiated in order for academics to gain recognition and ascend. Part of the 'purposeful' part of academic research can be seen to be tied up with questions about how doing particular work fits in with the positive impact that it will have on individual careers, disciplinary recognition and ascending in the institution. There are always various odd, messy, complex, stressful, challenging and emotionally draining things that occur during research processes and are central to these processes, which are glossed over or removed in the presentation of research outcomes. Academia is not divorced from social context and so, in a UK context, is not free from wider problems, such as racism (see Chapter 6 in Bhopal, 2018), sexism (see Chapter 4 in Bhopal, 2018) and bullying (Erickson et al, 2020), which clearly influence the ways in which research takes place, but will also not make their way into the ways in which academics write up their coherent narratives.

Research outputs tell narratives, and these narratives need editing to land successfully. It would be surprising to find researchers creating narratives that present what they have done, 'warts and all', or present themselves in the least forgiving light when their reputation and careers are at stake. This isn't to say that there aren't academic whistleblowers or those who speak out in ways that are detrimental to their careers, or that there aren't ways of doing and talking about research that make an effort to engage with what is often glossed over in outputs, or also that there aren't various mechanisms in academic writing that have been developed to check that academics are not achieving their ends via deceit or evasion. I am simply drawing attention to the ways in which the creation of narratives for particular audiences always involves considering what that audience values, and how they might react. Perhaps here, I am talking less about 'cunning' in the sense of 'Skill employed in a secret or underhand manner, or for purposes of deceit' – provided by the *Oxford English Dictionary* (OED) – and more about being 'canny', a term from Scots English and other northern varieties of English that relate to being wise, judicious, prudent and shrewd. Linear coherent research texts have significant value in various contexts as indicators of a researcher's productivity, impact, reach, knowledge creation, research significance and so on which have an impact on how well they are able to negotiate having a career.

Conclusion

Academic research helps us learn about the world and help others, but we need to be aware that much of what is involved in academic research relates to the creation of artifice. Research and theory involve simplification. The single epistemology. The drawing of boundaries around what is 'data' and what is extra. The article format. The creation of an argument. The choosing of elements to include in analysis. Considerations relating to how to handle subjectivity and objectivity. Learning how to negotiate these elements are part of learning how to apply research methods and, through processes of crafting and cunning, create texts that have currency with particular audiences. The different ways of interpreting the 'Shandy Bass Incident' were also tied to different identities. Steve as an artist. Richard as a historian of ideas. Kate as a literacy studies researcher. Johan as a philosopher. Each of these identities and interpretations emerged from varied biographical trajectories and privileged access to educational institutions (for example, art school or university education). In addition, these interpretations and identities influenced our professional and personal interactions as a group of people. Negotiating institutional norms and values is essential when attempting to leverage the resources of universities and funding bodies in the service of others. But there is an issue if we mistake the ways in which the work of academic researchers can be promoted for the complex work that is actually involved in responding to the needs of communities. There is also a sense that academic researchers are being positioned as 'up to this task' when the work presented in this book challenges whether this is the case.

To summarize, this is what I have learned about academic research, through the process of being involved in collaborative research:

- Research always emerges from specific social and cultural contexts (in other words, the institution is not divorced from the politics of everyday life).
- Research can help us think about the world, but it is also tied up with the ways in which academic systems shape how people can be seen to participate in academia, as well as institutional and individual agendas.
- Access to institutionally situated ways of doing academic research are unequally distributed, because access to the institution is unequal.
- Research involves a process of artifice, whereby through craft, cunning and construction, events and materials get 'worked upon' in order to create academic texts that speak to particular audiences.
- These processes privilege certain narratives and often obscure many other core elements on which the research activity depends.
- The emergent unfolding of everyday life is hugely complex, and so becomes simplified through research processes.
  - Showing that you can negotiate these processes is part of 'being an academic'. So abandoning them can cause personal conflict.

○ It can also potentially be easier to 'let go' of a disciplinary identity, if you already have a stable career and academic identity.
- The application of theoretical and methodological tools to create interpretations and narratives are tied up with different biographies, positionalities, access to educational resources, politics, preferences, identities, and professional and personal relationships.

The practices involved in doing academic research are involved in processes of accruing and maintaining prestige, heritages of ways of doing research influence what is seen as significant, and research is tied up with individual biographies and identities. When considering collaboration, it is interesting to consider the agendas and interests of the parties involved, and academics are working in an institutionally situated context with various cultural norms, pressures and politics. This raises questions about to what extent the 'academic career' is viewed more highly (by systems and actors in the institutional context) than the critical enterprise academic researchers are purportedly involved in furthering. The 'Shandy Bass Incident' was concerned with the experience of fieldwork and the question of what counts as research data. It has prompted a conversation about where we draw the line in research encounters.

4

Worthiness

Lalitha Vasudevan

Rashaad (youth researcher):	Tell me why I should believe you? Why do you care?
Lalitha:	What you have to say is worth my time and attention and I want to create space to listen to what you care about and change the direction of research, if we need to, so that the project is *ours*, not just mine.

Our life is an apprenticeship to the truth that around every circle another can be drawn; that there is no end in nature, but every end is a beginning; that there is always another dawn risen on mid-noon, and under every deep a lower deep opens.

Emerson, *Circles* (1903, p 301)

On the cover of the recently published children's book *I Am Every Good Thing*, written by Derrick Barnes and illustrated by Gordon James (2020), a young Black child, a boy of perhaps eight or nine years, is standing tall wearing a backpack and crossing his arms in front of him. The pages are filled with text that echoes with affirmations told in the first-person perspective, as the omniscient narrator is metaphorically transported into scenes of joy, enquiry and resilience – outer space, a baseball field, in the arms of family. On one page is the following text:

I am a brother,
a son,
a nephew,

a favourite cousin,
a grandson.
I am a friend.
I am real.

The accompanying image is of a boy – he may be the same as the child on the cover – hugging a smiling Black girl, whose hair is adorned with a pink ribbon headband and gathered in a high ponytail.

The dedication page at the front of the book lists the names of Black boys who were killed by gun violence, a few names I recognized and a few I looked up on the internet to learn more details about their lives. The book serves as both a *de facto* memoriam to the countless lives ended by violence, and a fierce call to action conveyed through the gentle rhythm of poetic fieriness.

The book concludes with an image of a young boy, no more than four or five years old, who is smiling and looking slightly up. The last line holds a world of wisdom and hurt: 'I am worthy to be loved.' These are weighty words. They serve as a reminder that the very idea of worthiness is contested regularly and that recognition or denial of worthiness is embedded in actions, words and practices.

Worthiness, as we proposed earlier, is a necessary precursor to empathy, which is vital for the type of collaborative and participatory co-production research we sought to pursue in the 'Reimagining Futures' project. The work of organizing research projects to create conditions for more than the mere inclusion of multiple voices in the research can be seemingly inherent in research pursued *with* (rather than *on* or *about*, for example) communities, organizations, young people, or some combination thereabouts. But the qualities of polyvocality and inclusion are not a given and need to be intentionally cultivated. Marion Dadds (2008), writing about empathy in practitioner research, argues that this type of enquiry – that is open to transformation in purpose and practice in response to the people involved – is equally valid: 'Related to the growth of empathy is the enhancement of interpersonal understanding and compassion. Research that is high in empathetic validity contributes to positive human relationships and well-being. It brings about new personal and interpersonal understanding that touches and changes hearts as well as minds' (p 280). But empathy is not automatic, it needs to be nurtured, coaxed forth and reinforced. Changing hearts and minds through research is not mere aphorism, but a mission embedded in research that is driven by a desire to change outcomes in people's lives. For our research team, the guiding light was the implicit pledge we made to pursue research activities and knowledge in a manner that resonated with, rather than detracted from, the participants in our projects. We began with a view that the communities and individuals with whom we worked and learned held an inherent worthiness, specifically to inform and shape the knowledge that would be produced and shared about their lives.

An empathic orientation in research can lay the groundwork for collaboration that allows all participants to flourish and does not advance only researchers' interests and goals. However, if we follow the logic that research produces knowledge about the way the world works, then we must also recognize that some of the research conducted about communities is premised upon beliefs that serve a narrative of unworthiness. That is, 'deficit' hovers as an immutable condition that is viewed as endemic to people and communities whose voices are often lacking or only cursorily included in the research about them. As a consequence, young people are increasingly posing questions to teachers, law enforcement and politicians about what their lives are worth – perhaps most vividly captured in the swell of the Black Lives Matters movement – and not always rhetorically, in classrooms, in community settings and throughout the vastness of virtual and digital spaces.

The 'Reimagining Futures' project likewise held a commitment to harness participatory approaches to research as a way to enact an orientation of inherent worthiness of the young people and the adult staff who cared for them; thus, our work was similarly shaped as research, but with a commitment to create conditions for worthiness to be engaged as a methodological north star. This project grew out of shared interests and curiosities, and expands and contributes to the legacies of participatory research with youth (Nygreen et al, 2006; Mirra et al, 2015), media and arts-infused research in out-of-school settings (Goodman, 2003; Soep and Chavez, 2005), ethnographies of youth and belonging (Abu El-Haj, 2009; Patel, 2013), and critical research on adolescents, literacies and justice (Moje, 2000; Winn, 2015; Fine, 2016).

Persistence of precarity

As the year 2020 melted into 2021, metaphors of the world on fire seemed all too apt, particularly as video of wildfires tearing through California echoed similar scenes in Australia less than a year prior. Grafted onto that backdrop, specifically in the US, the summer of 2020 was one marked by protests against police brutality that gave rise to a renewed national reckoning with race and the persistent legacy of slavery in the country, that is borne out most explicitly in the disproportionate rates of mass incarceration of Black people and hyper-surveillance in predominantly Black neighbourhoods. Coupled with the severe health, economic and social impacts of the COVID-19 pandemic, this reality was entangled with an intense US presidential election season, which was fraught, and the tensions resulting from it culminated in a near-insurrection at the US Capitol on 6 January 2021. This incomplete and frustratingly abbreviated summary of the 18 months during which time we wrote the majority of this book attempts to capture something about the quality of the air we breathed in when revisiting moments from projects in the near and not-so-near past.

When the story of Grace (pseudonym), a young woman living in Michigan in the US, rose to the top of trending news topics in the summer of 2020, the entanglement of justice, punishment and education was foregrounded (Cohen, 2020).[1] My 'Reimagining Futures' colleagues and I spent several conversations during the writing of this book reflecting on the connections between Grace's experience with the justice system and those of the young people we worked with over the course of a decade.

On its surface, Grace's is a story about a teenager being punished harshly for not doing her online homework during a pandemic. Yes and no. To say 'Yes' would be too simple. 'Yes' would mean that the judge set out to punish a 15-year-old Black girl for not doing her homework. 'No' is more complicated. No opens up the many missteps that led to Grace's detention in the first place. If the judge was earnestly concerned about Grace's well-being and wanted to give her an opportunity to "follow through and finish something", how did she imagine that this objective would come to fruition more effectively *inside* a detention facility rather than at home?

Is non-completion of online homework the reason the judge gives for initially removing Grace from her home – during a pandemic – and then keeping her incarcerated? Presumably. But missing from the interpretations of this story, and the judge's remarks, is context. That is, context from Grace's perspective. The judge herself revealed the context in her mind when she presided over the case, noting that she was "about to go over all the crap, all the negative, all the prior attempts at helping. I am going through it all". In this case, 'all' referred to the prior interactions Grace and her mother had had with the criminal justice system when law enforcement and lawyers were brought in, pleas from Grace's teachers were ignored, and an extreme response was issued for a situation that required deftness and care. In the process, context was disregarded.

The context is also COVID-19 and its highly infectious nature that caused schooling interruptions for nearly a billion children worldwide. The context is trauma from having the stable structure of a school day schedule pulled out from under you. The context is the stress of uncertainty, anxiety and inability to know how or when to ask for help. The context is a parent struggling to navigate her relationship with her teenage daughter. The context is a Black teenager living out her adolescence under the scrutiny of multiple systems

---

[1] In July of 2020, ProPublica journalist Jodi Cohen published a story about Grace, a 15-year-old Michigan high school student who was sent to juvenile detention for not completing her homework. That story gained traction quickly, sparked a #FreeGrace campaign across social media platforms, and updates to Grace's story were posted as Cohen and the ProPublica team continued to follow the story. ProPublica has devoted a separate page to stories about Grace, including a recent article about limitations on the use of restraints on young people in the justice system: www.propublica.org/series/grace.

of surveillance that have been shown to disproportionately detain and punish Black youth at a far greater pace than their peers.

Our conversations as a team, which had spent over a decade working together with court-involved teenagers, led us to ponder whether in a system besieged with limited resources and antiquated understandings about adolescents, Grace could be viewed as being worthy of an alternative reality than the one initially proposed for her by the presiding judge. Whereas the judge viewed her as a risk and a threat, Grace's circumstances could have been alternatively regarded as normative and deserving of support rather than punishment.

As some scholars and poets and other observers of everyday life have pointed out with a bevy of evidence and gripping prose, a confluence of life-threatening forces is commonplace for many people around the world. The unravelling of social safety nets as a result of the ongoing pandemic also serves to highlight the precarity of their infrastructure.

The work of 'Reimagining Futures' (at Voices) and its antecedent, the 'Education In-Between' project (at Journeys), unfolded in these persistent precarious social contexts, that of a social system whose origin is both celebrated and deeply flawed. Today, the juvenile justice system in the US operates as a punitive system, one that is premised upon beliefs and assumptions about youths' propensity for crime, their risk factors, and their contexts of home and community. What was envisioned as a boon to the well-being of children quickly morphed into a view of adjudication as punishment.

The example of Grace underscores the chasmic gap in understanding between the lived and imagined lives of youth and the adults who control the institution that govern their everyday lives. Nancy Lesko (2001) frames it this way:

> Static ideas about youth have helped to keep in place a range of assumptions and actions in and out of secondary schools. For example, since adolescents have raging hormones, they cannot be expected to do sustained and critical thinking, reason many educators. Since adolescents are immature, they cannot be given substantive responsibilities in school, at work or at home. (Lesko, 2001, pp 189–190)

The chaotic swirl of everyday stressors was not uncommon for many of the young people who walked in and through the spaces of Journeys and Voices. They carried heaviness in and on their bodies as they traversed hallways, building lobbies, elevators and stairs to make their way into the programme space. We noted that these young people, some as young as 11, would enter the room where we eventually convened, with their backs slightly hunched over, their necks somewhat lowered and their faces temporarily expressionless. The quotidian challenges weighed heavily on them similar

82

to the ways that global forms of turmoil are captivating the mental and embodied energies of many of us now.

We also witnessed the transformation out of this heaviness that was brought about through playful interactions with the staff. David Hansen (2018) writes, 'bearing witness is at once an ethical, aesthetic, and epistemic orientation' (p 23) and this description aptly reflects the ways that each of us on the research team held open our sensibilities and roles while engaging with the young people and staff at Voices.

In this chapter, I enact our collective witnessing as a mode of story-making and storytelling. Put another way, I aim to call attention to the rising and falling action of the narratives that comprise the collaborative research at Journeys and Voices. It was by bearing witness and, through that witnessing, creating conditions for young people to make themselves visible and known, that we carried out our work with these programmes. That work entailed ethnographic documentation, participatory research and pedagogical practices, in part, but it also was embodied/held/carried out in the work of co-producing research by cultivating relationships, co-creating our collective space with the youth and staff at the programmes, and inhabiting a space of unknowing. Elsewhere I have described unknowing as 'an act of dwelling in the imaginative space between declarative acts of knowing and not knowing; an invitation to wrest our modes of enquiry and our beings away from the clutches of finite definitions of knowledge and instead rest our endeavours in the beauty of myriad ways of knowing' (Vasudevan, 2011a, p 1157).

When we have shared stories at conferences about our collaborative research as part of the 'Reimagining Futures' project or the 'Education In-Between' project, audiences often pose questions about the inherent pedagogical ethos or approach in our ways of working with youth. For us, the answer was quite simple: we began with worthiness. But the simplicity of this realization is something that we have only recently come to articulate. Thus, the stories shared here offer a glimpse into that emergent insight – of presumed worthiness – and elucidate further what such an orientation looked like at various moments of practice in our projects.

There is no running without walking (except when there is)

The 'Reimagining Futures' project did not simply emerge into existence, but instead grew out organically from prior relationships. The origin story for what came to be the 'Reimagining Futures' project is partly rooted in my history as a teacher in the city of Philadelphia in the US. For a few years, I was part of the teaching team at an organization that provided educational programming for teenagers, aged 16–20, who were unable to return to their schools as a result of an arrest. They were mandated to attend the adjudicated youth programme that I was folded into as a teacher, despite my having no history of working in that context. I have discussed this experience elsewhere (Vasudevan, 2006), and simply note here that long before actively embracing ethnography as a way of being, humility in the face of all that I *didn't* know carried me as a teacher. Unknowing, at that time, was a pursuit borne of a need for survival rather than a well-planned-out pedagogical strategy. And for much of my time there, I learned alongside the young men and women who, at the time, were barely younger than I was when I began teaching.

I was nearly a decade older when I first learned of Journeys, an alternative to incarceration programme in New York City. The introduction to Journeys, an organization that would ultimately become a research home for 15 years, unfolded as a series of emails, first with my university colleague and then with a teacher she had introduced me to via email. Miles, the teacher who went by his last name, invited me to visit Journeys, located on the 12th floor of a high-rise building in southern Manhattan that was occupied by legal services, social service offices and courtrooms. (That building was renovated several years ago and is now a luxury apartments residential building.)

The entrance was located under a long stretch of scaffolding that lingered overhead for many years. I stood behind about a dozen people who were in line ahead of me and waited my turn to go through one of two metal detectors that stood at the entryway. A few weeks later, when I received my volunteer identification card, I was able to bypass the security check altogether. Each instance of cavalierly rushing around the metal detector gave me pause as I witnessed the removal and subsequent retrieval of belts, watches and other accoutrements that had, in some instances, been painstakingly coordinated and affixed as part of an outfit, but that needed to be removed and hastily reapplied as part of this ritual that has become an all too familiar part of social institutions in the US.

I was originally interested in how the youth at Journeys were navigating their education as they were court-mandated to attend this programme. My prior work as a teacher in a Philadelphia programme provided some foundation for productive unknowing about the lives of the youth at Journeys. That is, while I wouldn't presume to *know* the specifics of what the young people at Journeys were experiencing on a daily basis, either at home or at the programme, I came to observe the ways that interrupted schooling and its concomitant challenges of institutional labelling, hyperpolicing in the communities where many of the youth lived, neighbourhood tensions and unstable housing adversely affected young people and, by extension, their families.

Each of the roles I played necessitated different practices and, because of that, the relationships I formed with the youth participants, educators and teaching artists, counsellors, case managers, directors and support staff were also varied and multilayered. For a time, with my identification card in hand, I was welcomed into the organization as a partner – not quite *of* the organization, and not an outsider, but occupying a liminal space in between the dreaded insider/outsider dichotomy. This was not an easy road, nor was it simply achieved, for my naïveté about the transparency of my motivations and my ways of being as a researcher were bound up in how my very presence was read by people at Journeys.

One such role was that of documenter for the Theater Project that was started by two teachers at Journeys. EJ participated in one cycle and, after showing a keen interest in the fieldnotes and visual documentation my research assistant and I were producing, asked if he could also help to document the programme. I heartily agreed and invited him as an author on the collaborative research blog we had started for the new project. EJ flourished in this role, as both ethnographer – a word he instantly took a shine to – and as a mentor who readily gave feedback to the participants in the second iteration of the Theater Project. EJ painstakingly took descriptive notes, then audio-recorded additional notes into the digital voice recorder he borrowed from me, and was the most prolific contributor to the group research blog. When Kristine (who later became deeply involved in 'Reimagining Futures' and whose interlude is included earlier in this book) joined me in documenting the Theater Project, EJ served as a guide and apprenticed her understanding of the nuances of the project. It is important to note here that the 'participatory' aspects of our collaborative and participatory project were not predetermined, had not been outlined in a research proposal, and were not pre-approved by any entity. Yet, openness to the unpredictable places to which collaborative relationships would travel was an important aspect of recognizing worthiness. The unknowing mattered, and in its mattering, multiple new paths were forged and collaboratively explored through the use of enquiry methods that were commonplace (for example, fieldnotes, recorded conversations) and those that naturally unfolded from our collaboration (a spontaneously scripted dialogue, a music video, collages and other forms of art-making).

At the same time that we were documenting the Theater Project, I was asked by one of EJ's teachers to tutor and help him pass two sections of the General Educational Development (GED; high school equivalency exam). EJ and I would meet at various places around the city, including several meetings at my office on campus. He liked coming to campus so much that for a while, he was a regular presence in our department suite at the university where I work, with access to our project cubicle, a laptop and other resources. After passing his GED, EJ declared that he wanted to "do more" and he and another graduate of the Journeys programme, Joey, expressed a strong desire to provide support for the younger adolescents who had begun attending the newly opened Voices court-mandated afterschool programme by developing media-based and creative programming for them. Eric wanted to intervene before, as he put it, "they got into too much trouble".

Moments and realizations like this one resounded with something more than empathy – they called for recognition of worthiness. In the younger group, EJ and Joey didn't only see younger versions of themselves, they saw friends, siblings, children. They recognized something of the familiar, something worthy. Worthiness is a prerequisite for empathy to be called forth and enacted.

The work of reimagining young people's futures

'Reimagining Futures' was a project that was dreamed up by Kristine, EJ, Joey and me during an initial conversation about supporting younger adolescents. We envisioned the project as a workshop model that would allow us to facilitate different types of practices of communication, representation and knowing that included media creation and creative writing. The idea grew legs when we met with Vicky, then the education specialist at Voices, who saw the potential for an enrichment opportunity for the adolescents.

At the time, Voices had been in operation for nearly a year and Vicky knew us from her previous role as a case manager at Journeys who had been assigned to both EJ's and Joey's cases. She embraced our idea, as she did them, with her trademark love and care. Vicky met our proposal with enthusiasm and noted the need for programming at the still-growing Voices that was enriching and supportive of the young people who were court-mandated to attend Voices. In an interview several years later, Vicky reflected on that initial partnership and stated simply that she sought out partners who "loved [their] kids", referring to the young people who were court-mandated to attend the programme, a description that she ascribed to our project and our team.

"Do you love my kids?" This is an arresting turn of phrase that, while arranged as an interrogative, functioned more like a declarative statement. What factors shape how one might answer that question? Our response came in the form of prioritizing support for young people's well-being over whatever research proposal or design we may have originally submitted. It is worth noting there that this commitment did not neatly adhere to the expectations held by funders or by some journal publishers who often sought outcomes, measures of change, or indicators of predictability. Thus, rather than alter our commitment, my collaborators and I sought support from responsive funding streams, pursued publications in journals that embraced an expansive view of research, and, perhaps equally importantly, gathered institutional support to be able to continue this project for many years. In that time, nearly two dozen graduate students and half a dozen youth researchers contributed to and gained experience in being a part of this effort.

In our emerging collaboration with Voices, we were able to fuse programme development objectives with participatory ethnographic methods by carrying out EJ and Joey's vision. In doing so, we sought to design and cultivate spaces in which participants could make themselves visible through multimodal means, or what Jennifer Rowsell and Kate Pahl (2007) describe as identities that are sedimented across multimodal text production. However, the context of Voices held greater variability than its parent organization, Journeys (for example, myriad court dates, sudden changes in attendance requirements, school attendance record and obligations stemming from involvement in other systems of surveillance). This presented an insurmountable challenge to a few of the external community groups and programmes with which Vicky had tried to partner; they were accustomed to greater stability in participant attendance and inclination, and ultimately some either did not renew their programming at Voices or cut their workshop offerings short.

Worthiness can pry open *Chronos* (or chronological) time to allow the experience of *Kairos* (or how time is lived and experienced), an essential gift when nothing beyond attendance in a single 60-minute media workshop session, for example, could be guaranteed. Instead of seeking predictability and sustainability of programming, for instance, this orientation towards an embrace of *Kairos*, that became embedded in our approach as a research team, allowed us to foreground the goal of informing the practices in institutions like Voices that govern the lives of others – in this case, a programme that supports the lives of young people who are arrested and are court-mandated to attend an alternative to incarceration programme. Far from being a novel approach that we brought to Voices, our cultivated practice of 'making every moment count' reflected the patient urgency (Vasudevan, 2014) with which Vicky, Omar and the others at the organization expressed their commitment to young people. A programme evaluation or a survey administered at one point in time might have missed the time-bending that was a common occurrence at Voices, itself a practice borne out of a deep desire to move empathy from a distant concept into a lived practice. Seeing the young people as worthy of such efforts was foundational to that work.

Empathy in practice

Young people carry stories about worthiness with them. The moments when they share these stories live in words, silences, play and playful interactions, and engagement with the space around them. Collaborative research and co-production efforts have a role to play in scaffolding the practices of empathy to be revealed and communicated.

The question is how can methodology recognize and embody worthiness? Is it in the practice, orientation, activities, or all of these? In the 'Reimagining Futures' project, worthiness was baked into the project ethos, but this did not happen automatically. This commitment travelled across three sites, multiple staff changes (although core staff members remained largely consistent for most of that time), evolutions in funding and other disruptions. We ultimately arrived at the realization that worthiness was an essential ingredient of our research practice. As researchers who pursue collaborative and co-produced research spaces, perhaps we should be asking, in what ways does our research and our partnership reflect a commitment to worthiness? Here, I offer two glimpses, or 'takes' (akin to the cinematographic notion of a version of a particular scene), of an interaction with Robert, whom I met at Journeys and who participated in nearly every aspect of the programme. He attended GED classes, participated in the Theater Project, excelled in the internship programme, and signed up for the digital media course I offered. The first take is very close to the quick jottings I made when the interaction with Robert first took place, over a decade ago. The second take was written recently, after reflecting on this moment, reviewing fieldnotes, and having conversations with former 'Reimagining Futures' project members.

Robert

Take 1

Robert and I find ourselves in step as we walk with several other youth participants and Augustin (a teacher at Journeys) to the nearby art gallery in the SoHo neighbourhood in New York City. We pause in a side street while a few of the young men go into a bodega to purchase soda. It is springtime. Robert is wearing a tan, lightweight jacket. We are discussing another Journeys participant who was no longer in the programme. Robert looks up at me, squinting to avoid the sun getting into his eyes, and tells me, "When I was arrested, that was the first time I had a gun in my hands." I don't immediately respond. "Do you believe me?" he asks. I tell him that I have no reason not to believe him. He continues and says that "Most people don't believe me." Robert's giggle is high-pitched, distinct and infectious. He is slightly above average in height, had a stocky build, a round face, close-cut hair and dark brown skin.

Take 2

Robert was perched on a red brick stump that jutted out from the side of a store. He and I were waiting for several of his classmates and their teacher who had gone into the bodega around the corner to purchase something to drink. Robert shielded his eyes as we talked about Joey, who had graduated from the alternative to incarceration programme where Robert was still enrolled and would remain for another few months until the programme's completion. The sun was bright and remarkably warm for an early spring day, causing Robert to cast his eyes downward in between speech turns to avoid the solar glare. Around us, people walked by at different speeds of haste, which was a feat on the grey cobblestone side street where I stood and he sat.

Several seconds of easy quiet passed between us as we regarded the mid-morning passers-by, most carrying packages or bags, and nearly all wearing headphones or interacting with their smartphones.

Robert spoke, "When I was arrested, that was the first time I had a gun in my hands."

Or perhaps he said, "The first time I had a gun in my hands was when I was arrested."

I wasn't taking notes right then, and I didn't want to interrupt the moment by pausing to bring out my notebook or turn on a recording device. So the sequence of words ping-ponged in my head until I was able to capture it down later, while we started walking again and I could make my jottings.

A few seconds later, Robert followed this non-sequitur with a sobering question, "Do you believe me?"

His face was earnest, searching and lit by an ever-increasing intensity beaming down from the bright sky.

Another few seconds passed while I digested what he was asking me. I clumsily said that I had no reason to doubt what he was telling me. I didn't add that I started at a place of trust with him and his peers whom I had met in the programme. But his question foretold what he confirmed with his next utterance, "Most people don't believe me."

Prior to his reflection on his arrest, Robert and I had not been discussing the charges against him. In fact, the tacit understanding that I had with the organization is that neither I nor anyone on my team would bring up the young people's arrest records out of a need to protect their right to privacy. Still, the youth would seek out opportunities to think and talk through moments that made no sense to them. For example, during a digital media workshop I was teaching, Joey started to reflect on how he felt when a guidance counsellor told him that he should just drop out of high school because graduating would be too much of a Sisyphean endeavour. He joked that perhaps she was to blame for him turning his attention to less socially beneficial activities and then he resumed editing a video poem that he would ultimately spend ten weeks developing. Such moments of sharing were not rare when we gathered in classroom spaces in the programme suite.

At the moment of that exchange, I didn't know quite what to make of the fact that we were having a fairly vulnerable exchange on the street – albeit more a wide alley than one of the busier city streets – in the midst of a trendy New York neighbourhood, replete with galleries, popular restaurants, boutiques, and ample opportunities for window shopping and people watching. Now, many years later, the irregularity of the moment strikes me as relationally and spatially significant.

The echoes of these same corners and cobblestones call back from a different time. They linger and hover around as we continue our conversation. They have multiplied in this retelling of the story, etched with the memories I have of the same streets I trod, sometimes with and often without the young people from Journeys.

It is only with the distance of time and space that I could recognize that in that moment I was bearing witness, which is an orientation Hansen (2021) describes as connoting 'embeddedness in the world. It points to how a person turns (i.e. orients) body and soul in order to perceive and listen well. Perceiving is deeper than merely seeing, and listening is more encompassing that merely hearing. They are gestalts: whole experiences' (p 62). This is what Vicky and several of her colleagues at Voices and their counterparts at Journeys did every day as part of their practice of valuing young people as worthy of their care, attention and efforts. Empathy is subtle and can be practised readily when a reservoir of worthiness is available. Building and maintaining that reservoir is one of the greatest gifts that we received through this collaboration.

A note on institutional worthiness

Articulating the concept of worthiness has been a labyrinthine endeavour. How it is understood and enacted in the world, and how we enacted worthiness in our collaborative research is opaque. However, what is clear is that both the 'Reimagining Futures' research team and the staff at Voices and Journeys rested their pedagogical practice upon worthiness. From this foundation an expansive research practice can emerge including a stance of unknowing and enquiry as both an orientation and a goal. While worthiness is not among the metrics by which the organization operates or is measured, it is embedded in the founding of the Voices, as a place to support and serve young people.

Within contexts like Voices, researchers make choices about what questions to pursue and how to pursue them. A collaborative research endeavour, guided by co-production as an output, foregrounds *worthiness* as the foundation on which collaboration is built. Our ethnographic methodological orientation was shaped by Journeys and Voices into one that foregrounded practices of care (Valenzuela, 2017; Vasudevan et al, 2022). As Dadds (2008) described, it is possible to combine care and research by foregrounding empathy. The concept of worthiness as a key ingredient of research with young people brings this practice to the fore.

The formation of a group, with its concomitant components of shared characteristics and interests, brings forth the conditions for belonging. These conditions, as Mary Douglas (1986) argues, are not a given and whereas she is explicating the thinking that drives institutions, her astute observation about how an institution – even a collaborative research group – comes to matter enough to effect behavioural change is resonant with the phenomenon we observed when embarking on collaborative efforts together with Journeys and Voices. The organization had its goals, informed by numerous external factors (such as funding agencies, with their assessments and predetermined metrics of success, like attendance, test scores and rates of recidivism), the commitment of its staff and honouring the legacy of justice reform on which it was founded. Additionally, the programmes we worked with were focused on meeting reporting requirements, achieving compliance with standards of practice set by the government agencies to which they were tethered. Even so, while worthiness is not among the metrics by which the organization operates or is measured, it is embedded in the founding of the Voices, as a place to support and serve young people. Likewise, we had project goals that were framed in part by research questions and prior iterations of our collaborative efforts, research funding expectations and a desire to provide an experience that was value added for the young people as well as fulfilling the expectations of the organizations. These competing interests could have been paralysing if a foundation of mutual vulnerability and commitment to the worthiness of young people had not been established.

As the chapter has shown, there needs to be a fundamental shift towards valuing young people, a shift toward what I refer to here as 'worthiness', which then signals a necessary shift in research practice. Institutions have a role to play in reimagining how they encourage and support collaborative research whose outputs are not confined to reproducible findings, static or causal explanations for social phenomena, and the like. The implications for such a role are significant and include, but are not limited to, expanding the infrastructure in universities to encourage research that is free from outcomes (at least at the outset); lowering barriers for community partners to gain greater ownership over research practices and artefacts; cultivating audiences for this type of collaborative research in publications, curricula, media and other forums that hold currency across the different domains of research partnership. The road ahead for supporting unpredictable, collaborative and empathetic research that values young people and communities is certainly complicated, but the consequences of not doing so are far more dire.

INTERLUDE 4

Two

Jonathan May

Making photographs is a collaboration, and like all relational practices, the hardest part is to show up. To be there with the camera. To trust intuition and lift it to your eye. To peer through the prism, believe in what you saw and respond to the moment. It's a conversation with chance in the unfolding now. I have an idea of what the image will look like, but when the shutter is pressed my frame goes black as the mirror lifts to let the light spill in. I look through that mirror to set the frame, and then believe in what happens next.

Crowning in the birthing pool, my first glimpse of you was through a mirror held by a midwife, lit through the water like a cave dive. Within moments, you were pulled from the pool and passed to me; pale, moist, streaked with blood. I lifted you to my chest, the first time I had ever really held a baby. Hot flesh, trembling limbs, shock of wet hair, paper thin fontanel. And I remember S telling me that the most important thing to do as a father is to show up.

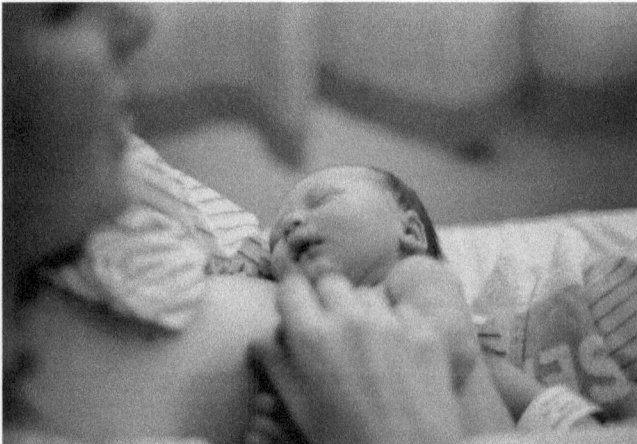

You began as two. You and your mum intermingled sharing cells, blood, oxygen, glucose, nutrients. And I existed in orbit; sometimes pulled close and often carried farther than I liked. I'd run baths and rub feet, brew tea and clean up, occasionally catch you pushing to me and an outside you'd yet to conceive of. And even after you were drawn from that water, you two were still half connected, gnaws of hunger pulling you back to your mum and her nourishment. She loved you with her whole body as your beings still mingled. And me once again on my orbit, leaving and returning while you fed and grew. On that orbit, I learned a bit about you and a bit about me; how to say goodbye and how to meet you again. That first year was about you, but here in your second year, we learn about us.

I have photographed you every day of your life that I can remember, but making photographs of us is our first collaboration. Where once I was separate and behind the lens, now I am before it with you. Held together within its fixed field of view, the photographer and subject become a father and daughter. Our hierarchy is levelled and tensions laid bare for the reciprocity of the moment to play out; where you react to me and I respond in turn, rarely in ways I expect.

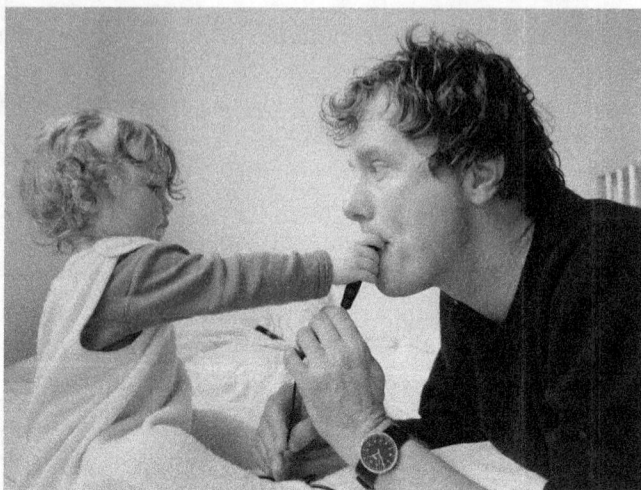

My hand in each photograph is inescapable, so I'm careful not to press it too heavily. I want you to be yourself; to give you the space to perform your own behaviour, and to explore what our behaviour is. A space full of curiosity and misunderstanding within which come sublime moments of discovery. Sublime because they are discovered together through a mess of confused intentions and shifting perspectives. It's a commitment I want to make to you; a principled and contradictory commitment to push, to explore, and to compromise and to be changed.

I know that right now you don't know we are making photographs. You won't remember when we started and one day it will be as natural as egg and soldiers. By the time you read these words, these photographs will have always been there with us. Proof perhaps, of where our relationship started.

Because in photography there are the images we create, and the moments in which we make them, and then there is the practice of making them.

The reality of the photographs we make is that I am not in control, and neither are you. I anticipate an image, and I position the camera, but I don't know what moment will follow. I set the frame, but the rest is our improvisation. The shutter opens to chance as we transform whatever image was in my head that compelled me to reach for the camera. The idea of the image is important because it's what gets me to take the camera out, but the best photographs emerge from the unfolding moments that follow.

The idea is the belief that gets me reaching for the camera, but once the camera is out, I never know what the picture is going to be. You have to have the confidence to not know, and go forward anyway. And if there is one thing I would like you to learn from the photographs we made together, it is exactly that; to go forward in the courage of your love. Because I'm not going to be here to remind you every time.

5

Enchantment

Richard Steadman-Jones

Lalitha:	How does this work alter one's orientation towards the objects that one investigates?
Richard:	You stop feeling as though you are developing expertise on what you are investigating and instead, come to see it as extraordinary and magical.

No one can remember quite where the idea for the 'Language as Talisman' project came from (see Appendix). At the time, we were talking with the headteacher of a local primary school. They had had a difficult time with the Office for Standards in Education (OFSTED), the body which inspects schools in the UK and produces public reports on these inspections. In particular, the inspectors had been critical of the use of dialect by teaching assistants and children in the classroom. They saw it as their role to reinforce the standard language ideology, which characterizes non-standard speech as aberrant, inferior and inappropriate. Somewhat outraged at this approach, we wanted to develop a project in which we worked with local people, children included, to develop a nuanced account of their own language use – one that did not dismiss their speech as substandard (Escott and Pahl, 2017; Hyatt et al, 2017). We felt that ideas from the discipline of Linguistics might be helpful in enabling children to develop a reflective approach to their use of language. For example, Communication Accommodation Theory (Coupland and Giles, 1988) provides an account of how speakers vary their speech in response to different interlocutors, and we saw this as offering a way to think about the different varieties that the children used and the different contexts in which they used them. At the same time, we did not want the project to become too burdened with technical concepts, and at some point, someone came up with the idea of locating the work in an exploration of the 'magical' properties of language – instances of language use, we felt, could be compared to talismans. A talisman, as defined by the *Oxford English Dictionary* (OED), is:

> A stone, ring, or other object engraven with figures or characters, to which are attributed the occult powers of the planetary influences and celestial configurations under which it was made; usually worn as an amulet to avert evil from or bring fortune to the wearer; also medicinally used to impart healing virtue; hence, any object held to be endowed with magic virtue; a charm.

Our feeling was that language could be talismanic in the sense that it could protect the user from malign influences – 'avert evil or bring fortune'. For example, to accommodate one's usage to a particular speaker is to signal solidarity and sympathy in a way that might materially affect one's relationship with them and, hence, the way in which they treated one. To think in this way was to undermine the idea of language as a site in which young people might be tested and found wanting, characterizing it rather as a source of security, power or influence. And we thought that the comparison with a talisman, amulet or charm would be an accessible way to explore these properties of language at quite a sophisticated level.

With these ideas in mind, we began to explore the metaphor of 'language as talisman' in greater detail and I (Richard) produced a film examining six ways in which language could be thought of as talismanic. The film has the quality of bricolage. It draws upon texts that happened to be on my bookshelves but which open up the meaning of the comparison in helpful or provocative ways. Here is the first section.

Power and protection

In 2011, Alistair Kwan published a paper on what he calls 'talismanic technology' in the thought of the Renaissance occultists Tycho Brahe from Denmark, John Dee from London and Cornelius Agrippa from Cologne in Germany. In 1533 Agrippa, the third of these great 16th-century magicians, published a work in Latin under the title *De occulta philosophia libri tres* (*Three Books of Occult Philosophy*). Talismanic technology is a central theme of the work. Kwan (2011) puts it like this:

> Agrippa begins with an overview of talismans, citing numerous references from antiquity, the middle ages, and times more recent. ... [He] also suggests making images to draw down beneficent celestial influences and offers an overflowing thesaurus of information whence to design them. He gives for instance numbers and magic squares and

obscure symbols associated with each planet and also stones, fragrances, places, foods, and colours. (p 107)

The English edition of Agrippa's text appeared in the middle of the 17th century. One section describes a number of ways in which the magician can 'draw down beneficent celestial influences' from the moon and it includes a number of 'obscure symbols' associated with particular aspects of this esoteric power. Agrippa claims that material artefacts inscribed with these glyphs can offer the magician some very particular benefits: 'This fortunate Moon being engraven on Silver, renders the bearer thereof grateful, amiable, pleasant, cheerful, honoured, removing all malice, and ill will.' What is more, 'It causeth security in a journey, increase of riches, and health of body, drives away enemies and other evil things from what place thou pleasest' (Agrippa, 1651, p 242).

How did thinkers like Agrippa believe that talismans worked? 'Talismanic technology', according to Kwan, relies on the 'mechanism' that mediates the 'influences' of bodies such as the moon. 'We moderns', he adds, 'might think of them as being a bit like a radio receiver'. (In the film an image now appears of a crystal radio made by hand in the middle of the 1930s.) 'Radio receivers (and talismans) receive all frequencies all the time, but are tuned to respond far more strongly to the narrow band occupied by one station (planet), while suppressing signals from all the rest' (Kwan, 2011, p 103). So, while we cannot hold back all the music we dislike, we can set our radio receiver to respond more strongly to some other more desirable signal.

And so, according to Kwan, the Renaissance model of the talismanic saw the individual as surrounded by flows of power: complex, constant and capable of both good and evil. The point of constructing a talisman was to draw flows of beneficent power towards you or, conversely, to ward off influences that might do you harm. For Agrippa, then, the talisman provided a means of reordering the world around the self and building a protected space.

This glimpse of Renaissance magic offers hints of what it might mean to think of language as having talismanic properties. It would involve seeing words as a means of protecting the self from harm and intervening in external flows of power: not the flows of power that Agrippa saw as emanating from heavenly bodies but social power, political power, the power encountered in the everyday experience of life.

★★★

This is an example of a situation in which one is carried along by one's own writing. The comparison with Agrippa's lunar talismans lends to ordinary language an aura of enchantment. It is not simply that the metaphor explains the functioning of language on a conceptual level. It turns language into a site of magic and it turns speakers – the children from the primary school among them – into magicians. This is particularly productive if one thinks of the intervention of bodies such as OFSTED in terms of the disenchantment of everyday language use. The concept of disenchantment, as we said in Chapter 2, comes from the work of Max Weber and it constitutes a way of characterizing the particular features of modern power and governance. Disenchantment involves the conceptualization of the whole of life as knowable, calculable and, hence, governable by means of bureaucratic systems. It involves the measurement and assessment of the exterior world in ways that render it susceptible to rational forms of control. A body such as OFSTED can be seen as a disenchanting force – a prototypical part of the bureaucratization of everyday life. However, the identification of disenchantment need not become a counsel of despair. For some critics, where there is disenchantment, there can also be re-enchantment. Jane Bennett (2001), for example, argues that life is lively with opportunities for enchantment and that enchantment has a particular moral quality. Her point is that for ethical living, it is not sufficient to have a particular moral code. One also has to have a disposition to follow that code and this disposition has an important affective dimension. Enchantment, she argues, can produce the kind of affective orientation that makes moral living possible. Hence the re-enchantment of language can be seen as wresting ordinary communication from a force that renders it susceptible to assessment in bureaucratic terms and as restoring to it its moral power. Language can be a site of wonder and this wonder can have significant ethical properties.

The script of my film continued by looking at other ways in which language could be seen as magical – ones to do with the power and creativity of tradition.

Tradition

Our second glimpse of the talismanic comes from the pages of *Religion and the Decline of Magic*, an important work about popular belief in England published by the historian Keith Thomas in the early 1970s. Early in the book Thomas (1971, p 33) talks about the use of talismans in late medieval Christianity: 'Theologians held that there was no superstition about wearing a piece of paper or metal inscribed with verses from the gospels or with the sign of the cross, provided no non-Christian symbols were also employed.' A very widespread example is the '*agnus dei* (in English, the Lamb of God) a small, wax cake originally made out of paschal candles', the candles used in the Easter service. The *agnus dei* bore the image of the lamb and flag, a motif that, for medieval viewers, represented a symbol of the risen Christ. Thomas adds that 'this was intended to serve as a defence against the assaults of the devil and as a preservative against thunder, lightning, fire, drowning, death in childbed, and similar dangers'.

The *agnus dei* is also the name of a prayer that beseeches the risen Christ to have mercy on the speaker:

> *Agnus Dei, qui tollis peccata mundi, miserere nobis.*
> *Agnus Dei, qui tollis peccata mundi, miserere nobis.*
> *Agnus Dei, qui tollis peccata mundi, dona nobis pacem.*
> [Lamb of God, who takes away the sins of the world, have mercy on us.
> Lamb of God, who takes away the sins of the world, have mercy on us.
> Lamb of God who takes away the sins of the world, grant us peace.]

Of course the *agnus dei* is not the only form of talisman to be found in the Christian tradition. People making journeys may carry medals of Saint Christopher, and the reason that Saint Christopher was designated the patron saint of travellers is because of the legend in which he carried the Christ child across a swollen river, staggering under the weight of sin that the divine infant had taken upon himself.

The talisman associated with Saint Bridget is a particular type of cross. It consists of a square with four arms emerging not from the centre of each side but as continuations of their edges. It is not a depiction of the saint herself but an image she is said to have created by weaving together rushes – a sign of the crucifixion made from the native plants of the Kildare countryside. But the history of Saint Bridget's cross seems to extend into the era before Christianity. It was once a pagan symbol associated with the *goddess* Bridget, an image from one tradition appropriated into another along with other properties of the original Celtic deity.

Thomas' text provides us with a glimpse of another talismanic tradition – not the celestial magic of the Renaissance but the popular beliefs of ordinary Christians throughout history. And other religious contexts offer similar traditions – in Islam, for example, it has sometimes been the practice to carry a talisman inscribed with a quotation from the Quran.[1] Again, this may offer hints as to what it means to see language as talismanic. Particular forms of talisman are rooted in particular traditions and in the same way

---

[1] See the Met Museum's 'Might and magic: the use of talismans in Islamic arms and armor', www.metmuseum.org/blogs/ruminations/2016/talismans-in-islamic-arms-and-armor and 'Amulets and talismans from the Islamic world', www.metmuseum.org/toah/hd/tali/hd_tali.htm.

the use of language is strongly influenced by the traditions with which people live. Language may in general offer a means of mediating flows of power and protecting the self from harm, but how this works in particular instances depends upon a history of practice. Sometimes uses of language draw strength from their conformity with tradition, and at other times, from a process of transformation that renders material from the tradition uniquely theirs.

★★★

Thus, the language use of the community members with whom we were working became subject to another layer of re-enchantment. Not only was the individual's use of language magical in the sense that it mediated flows of power and offered a species of protection; the process of re-enchantment extended to the community as a whole and to the history of its language use. Ways of speaking passed down through the generations – local varieties and the vernacular rhetoric of the group – can also be seen as invested with magical properties resistant to the rationalization of official agencies. And young people within the community were also initiating new traditions – within the local park a small skate park had become the focus of new forms of practice and new types of inscription. Particular ways of doing things and particular forms of speech were emerging as part of a collective mode of practice associated with the skate park. Language is, after all, collective, and its magical properties are embedded in the traditional practices of the group, both old and emergent (see Pahl, 2014).

As the film continued, it turned to the history of psychoanalysis.

Narrative

Our third glimpse of the talismanic comes from the work of Carl Jung, who finds the giving of a talisman to be a traditional motif in fairy tales. And this is important because for Jung, fairy tales offer clues about the collective unconscious, the shared store of archetypes that structure the human encounter with the world. Jung compares the lifecycle to a journey in which the young hero or heroine sets out from home, encounters obstacles and challenges, and returns finally stronger, wiser and more fully integrated as a human subject. Ideally, then, the end of life is a coming home and this can be suggested through the sign of the Zen circle, a path that returns to its point of origin and, in so doing, traces an image of completeness, perfection and simplicity.

Along the life journey, the hero encounters archetypal figures, aspects of the self that separate out to interact with the traveller in a variety of different ways. Among them is a benefactor both older and wiser than the hero – a figure who has important connections with the theme of our discussion.

Often [says Jung] the old [benefactor] in fairy tales ... gives the necessary talisman, the unexpected and improbable power to succeed, which is one of the peculiarities of the unified personality in good or bad alike. But the intervention of the old [benefactor] ... would seem to be equally indispensable since the conscious will by itself is hardly ever capable of uniting the personality to the point where it acquires this extraordinary ability to succeed. (Jung, 1991, p 88)

The nature of the talismanic object varies from story to story, and in a footnote, Jung mentions tales from different contexts: two stories from China, one in which the gift is a book of secret wisdom and another in which it consists of three miraculous dogs; another from the Balkans in which it is a flute that sets everyone dancing; one from Scandinavia in which it is a path-finding ball; and, finally, one in which a girl who is searching for her brothers is presented with a bundle of thread that rolls along the ground and takes her to them, this last story coming from the Baltic.

And so Jung's references to talismans place them very firmly in the context of narrative. The giving of a talisman is an important episode in the story of the self and it happens at the moment when the wandering hero or heroine most needs strength and resolve.

Again, this might lead us to think afresh about the talismanic properties of language and how they fit into the story of a person's life. Where does the ability to use language as a source of power originate? Under what conditions does it emerge? Is it encouraged by the intervention of some benefactor, and how is it related to the story of trials borne and conflicts faced? What is more, we might ask if narrative itself can be talismanic – if the telling of a story can itself bestow upon the teller 'the unexpected and improbable power to succeed', which Jung suggests 'is one of the peculiarities of the unified personality in good or bad alike'.

★★★

One of the participants in our project was a youth worker who often described his work in Jungian terms. He saw himself as an archetypal figure, encountering young people on their life journeys and helping them with their passage through his particular neck of the woods. Whether or not one subscribes to the Jungian psychoanalytic framework, to imagine oneself in these narrative terms is to reject – or hold at bay – a bureaucratized account of the work of the youth worker, hemmed in by policy documents and key performance indicators. It involved our colleague casting himself in a narrative where the young people who attended the youth centre were the heroes. Language of all kinds can be understood in narrative terms. Words and phrases are interwoven with the stories of individuals. We can recall when someone produced a particular utterance – the punchline of a joke can be something someone said. If language is magical as a result of its protective function and its place in the collectivity, it is also connected with stories (or myths), and the narrative that surrounds the language is in itself enchanting.

Finally, the film moved on to the talisman as a physical object.

Craft and materiality

The fourth and final glimpse that I shall offer comes from the work of Marcel Mauss, a French anthropologist who published his book *A General Theory of Magic* just two years into the 20th century. Like Jung with his fairy tales, Mauss (2001) drew upon descriptions of magic as practised by men and women right across the world. His aim was to identify some common threads to all these practices – to explain what constituted the category of magic as opposed to related ones, particularly science and religion.

Writing about the practices that are prototypical of the field of magic, Mauss (2001) emphasizes the importance of material substances in magical rites: 'The magician's shrine is a magic cauldron. ... Ingredients are chopped up, pounded, kneaded, diluted with liquids, made into scents, drinks, infusions, pastes, cakes, pressed into special shapes, formed into images' (p 66). He goes on to say that this kind of work is seen by the magician as bestowing a special quality on the materials themselves: 'They are drunk, eaten, kept as amulets, used in fumigations. This ... serves to provide them with a ritual character which contributes in no small way to the efficacy of magic' (p 66).

The emphasis, then, is on the talisman as a physical object that must be constructed from raw materials through a variety of ritual techniques. And Mauss goes on to list a wide range of such materials in his description of magical rites:

> Magicians prepare images from paste, clay, wax, honey, plaster, metal or papier-mâché, from papyrus or parchment, from sand or wood. The magician sculpts, models, paints, draws, embroiders, knits, weaves, engraves. He makes jewellery, marquetry, and heaven knows what else. These various activities provide him with ... his symbols. He makes gree-grees, scapulas, talismans, amulets, all of which should be seen as continuing rites. (2001, p 66)

And so Mauss directs our attention to two more aspects of our subject. A talisman may be the product of a tradition but it is also the result of someone's effort. It is, in short, the outcome of craft. And, what is more, it has a physical existence. It is made of something and what it is made of is significant to the person who comes to use it.

Language can also be the object of craft – worked, shaped, polished, remade. And both speech and writing have a material dimension. Sound waves disturb the air and writing requires a medium whether ink or paint or blood. Our account of language should include some sense both of its status as craft and its specific material forms. These are important to its users and they should be to us, too.

★★★

As part of the project, we produced a literature review (Escott and Pahl, 2012) covering relevant texts from a variety of intellectual traditions. The review highlights a number of texts that discuss the crafting of language. For example, in relation to David Barton and Uta Papen's (2010) book *The Anthropology of Writing*, we wrote, '[t]he book as a whole looks at writing acts as being embedded in social, cultural and even historical situations whereby even everyday writing is seen as an artisanal and therefore crafted process' (Escott and Pahl, 2012, p 16). Similarly, we said of Nikolas Coupland's (2007) book *Style*, '[t]he author's theory of style looks at the social context of linguistic forms and processes and how aspects of style can be seen to be chosen or crafted' (Escott and Pahl, 2012, p 21). Thus, there is a thread in Linguistics and Anthropology which centres on the ways in which everyday language – not just, for example, literary writing – is the object of craft, and we might see this in the polishing and improvement of a joke or story that becomes a little more artful each time it is produced. Language also has a material existence and the objects through which it is instantiated can themselves be associated with enchantment. A young man who attended the youth project produced a medal that had belonged to his grandfather (or perhaps his great-grandfather), and the medal called up stories – narratives of the war which were particularly potent within the community (see also Pahl and Rowsell, 2010).

Thus, the film came to an end.

Epilogue

Four glimpses of the talismanic and six properties of talismans:

Power
Protection
Tradition
Narrative
Craft
Materiality

So, after all this, what does the term 'talisman' offer us? It offers an image of the richness of language as it does its seemingly magical work in the constitution of human communities.

★★★

At the beginning of this chapter, I mentioned that our aim in the 'Language as Talisman' project was to use ideas from Linguistics (and Linguistic Anthropology) to enable community members to reflect upon their own language use. Initially, the comparison of language with talismanic artefacts was simply intended to provide a more accessible way to mediate this reflection. However, thinking more carefully about the comparison led to the idea that the project might constitute a way of re-enchanting language after the school's bruising encounter with the forces of disenchantment. One might wonder, at this point, whether there was a tension between the use of concepts from Linguistics and the overall project of re-enchanting language. After all, Linguistics is understood by its practitioners as the *science* of language and science is often implicated in the process of disenchantment, offering the means by which experience is rendered rationalized and calculable. However, this tension does not in practice seem to have arisen, and this is probably because it is a fundamental axiom of Linguistics that all varieties of language are equally valid and interesting. Linguists do not condemn speakers of regional varieties in the way that the representatives of OFSTED sometimes do. Hence the forms of scrutiny that were central to the OFSTED inspection – the checking of classroom talk against some notion of 'better' and 'worse' language use – have nothing to do with the ways in which linguists think about their object of study, and the 'science of language' can be an ally to the process of re-enchantment, finding objects of interest in the most everyday forms of communication.

When one writes a bid for funding, one cannot specify the re-enchantment of language as a projected outcome. Indeed, the more cynical reader might feel that the funding bodies are themselves agents of disenchantment with their emphasis on measurable outputs and their attendant systems of accounting. Yet, this was a vital element of what the 'Language as Talisman' project did. To return to Bennett (2001), the experience of locating magic and wonder in the everyday language use of participants constituted a kind of moral intervention – the production of an affective orientation that reinforced our commitment to the well-being and progress of the participants in the project. Perhaps the time will come when one *can* list re-enchantment as one of the outcomes of a proposed project, but until then, it must remain an 'unofficial' and quietly subversive constituent of research.

INTERLUDE 5

Demons

Richard Steadman-Jones

There is a figure in the history of Tibetan Buddhism called Milarepa. A legend has it that one day Milarepa went into his hut and found that it was full of demons. Disconcerted, he told the demons to go, but they just laughed at him and ransacked his cupboards for food. Nothing Milarepa said made any difference, so in the end he gave up berating them and sat down with them quietly, at which point, all but one of them vanished. Suddenly sure of what to do, Milarepa walked over to the one remaining demon and put his head into its mouth. When he did this, that final demon vanished too.

The story is about the power of vulnerability. It asks how vulnerable we dare to be.

Do we have the courage, rather than storming around and telling the demons to go, to sit with them, and even invite them to eat us up?

6

Embodiment

Lalitha Vasudevan

Richard:	The feelings that the projects generated seemed important. How can you capture them in the way that you document them?
Lalitha:	I like that question. I think this chapter is trying to capture affect through images (and sounds and other sensations, as well, through images), and highlights the feelings that echo through a project.

Making time

A social worker I interviewed for 'Reimagining Futures' described Omar, a mutual friend and colleague, as someone who "always made the time." The emphasis in her words suggested that '*making time*' signified the ability to stretch the affectively embodied properties of linear time. Omar had an unhurried manner and could seemingly hold time at bay. It is this friend's arm that is slung across a young man's shoulder as they listen to stories about African history and the secrets that artefacts hold. Standing together, the two of them travelled millennia in minutes, and their journey remains frozen in that moment.
(100)

Poetry, of the sort that is found in the natural world and only ever partially represented by the written word, is as close to a felt definition of embodiment as there is. Poets author poems that hold precision in their abstractions and can have a transportive quality that effects a change – however imperceptible – in how one receives the natural and material world thereafter.

Images, too, can transport, sometimes forcing the viewer to catch one's breath or laugh or gasp or weep. So might the feel of sand slipping through fingers, wet grass on toes, the first snow of the season.

Fractals of easily repeated forms are plentiful, captured imperfectly in image, like a photograph. But how to capture fractalized affect? The reverberations of laughter sparked by a glance or a word misheard by …

Dear Ed …
Dear Nicole …
Dear Yvette, Jose, Alfonso …

Dear Juan,
I once placed a hand on your shoulder. I didn't know what else to do when you told me your friend had been killed – shot – over the weekend. Like your face, the page of fractions below you was tear-stained, and as I reached over to hand you the book of poetry you seemed to gravitate to almost daily, you revealed that you had witnessed his death.

Dear Brite,
Your name sings in the key of your energy, even if you needed naps in class every day. How easy it would be to ignore or dismiss you, to read your fatigue as disruptive or disrespectful.

	Shifting away from a view of photography as a way of seeing to one that regards the temporal power of photography for making time, I revisited several sets of images I had ~~taken~~ made while travelling, studying, learning around the world. More often than not, I remembered where I was, what was happening, and what I was using to make the image – to create the photo. I recall exact moments the shutter-release on an SLR digital camera was momentarily depressed at that exact time. The lens of a Canon digital SLR lays in my left hand differently than my iPhone. (100)
… chance that gives rise to robust digression, complete with spontaneous soundtrack and references to movies, television, and other forms of popular culture. How is affect embodied in how affect is conveyed? The same sensorial categories can be ascribed to almost anything. *The air escaped the room when Malcolm walked through the doors, not with his usual bounciness, but instead with shoulders slumped and a bandage over his eye. Omar met him at the door and placed a hand on his shoulder and together they walked with lead feet towards the table in the centre of the room.* ★★★ *Leo walked up as close to the painting as he could without setting off the alarm. The soft, overhead light warmed his face as he studied the brushstrokes interspersed between and …*	Hi guys, how's it going? *Trey swayed gently as he spoke, initially speaking barely above a whisper. Then, with his voice breaking, amplified his words to talk about his 'second father' Omar.* Jay, it's good to see you again. Lester was here last time so he should be able to fill you in – go ahead Lester, you're taking over the class today. "Me?!" [laughter] *Trey stood in a sea of faces, hidden at first behind three rows of mourners who had come to pay their respects and witness final rites. This was the third memorial for Omar, and the third one that both of us had attended, along with many others.* Lydia is rolling out this large paper on the table for us because we are going to keep asking questions today and talk more about …*

The weight of the time-making tool has as much to do with the image being made as the activity or phenomenon that invites documentation. Inclination, opportunity, instrument and action converge and the image becomes the resin in which time is preserved. Some images remain timeless, while others are yoked with the conditions and context of their origins.

But, when is a photograph made? When I see a group of three boys outside the window of the car who are running and walking their bicycles and tires alongside traffic on a busy Kerala roadway? When one of them looks at me?
(100)

... over the top of the bits of magazine, newspaper, clippings of letters and hastily torn menus that had been pasted to the canvas.

★★★

Temporary railings that had been piled together against a concrete barrier clanged loudly in the afternoon air as first Ronald, followed by several others, vaulted over them. Their wide grins and squeals of laughter filled the video frame, jubilant and almost completely free.

★★★

Baritone, more a feeling than a sound, floated towards the hallway first. Not singing, but a story being shared. Laughter was intermittent as the baritone voice grew louder with each step. Ty was only 17, but spun stories like someone who had survived far more decades.

At work is the impact of remembrance –

... qualitative research. [laughter] "Miss! Am I going to interview someone? Can I interview LeBron?!" [laughter]

Once, while Omar and I were seated and talking in his narrow office, Trey knocked on the door. Wordlessly, Omar reached for a paper clothing bag with woven handles that had been placed behind the door and handed it to Trey, who nodded and mumbled "Thanks," casting his glance downward while he talked. Omar explained more after Trey left the door frame, presumably to go home – but it would turn out he was sitting in the classroom, waiting to talk to Omar again. "He's been going through some stuff with mom," is how Omar described Trey's current living situation: no longer welcome at home, staying with friends on their sofas, and continuing to show up at Voices even though he was no longer court-mandated to attend.

Upon first, second or third glimpse of the desk and chair that appeared in the middle of a field on Tory Island, just north of County Donegal in Ireland? Coming face to face with a mossy stone wall tucked away on a street in Nicosia, where stories from the past whisper loudly? Looking up just as the man in a red shirt appears in an opening of a Parisian building? Stumbling onto my grandmother's childhood neighbour in the village where she and 15 of her siblings were born? Finding the meaning of a theatrical production embodied in the cast's footwear? (100)

... for example, a pencil used by a child during an activity can snap that moment back into the present, not in a supernatural way but rather in the ways that the traces of form and function linger on long after any given moment of interaction. The same pencil might have been used to shade in a drawing, to gather and fasten a messy bun atop a child's head, or to coax a crumb out of the keys of a laptop keyboard. The pencil does not have a fixed history; the pencil embodies the history of many stories with which it intersected.

What about 'other material that hasn't been paid attention to and that we're trained not to pay attention to [emphasis added]? The theorist or the writer or the cultural producer is trying to say, "Well, what if you paid attention to this set of things? How would you deal with what's ...

Let's start with a little bit about the *why* of qualitative research and –
"But how are we going to know they're not lying? When we interview someone, what if they lie?"
That's a great question, Terrence. Sometimes you won't. And ... Ok, let's pause here and talk about truth. What are some things you know to be true? *Really* know to be true.

Omar ran into Trey almost a year after he had been discharged from the programme, at night, 'hanging out on the corner up to no good'. The circumstances of that encounter were less than ideal. Omar described the way he had implored Trey to come back to Voices – where he could always talk to him and Vicky – whose listening ear exuded empathy – and receive help. A few weeks later, Trey appeared at Voices again, consuming the snacks and small dinner-like meal that was always available and hesitantly socializing with ...

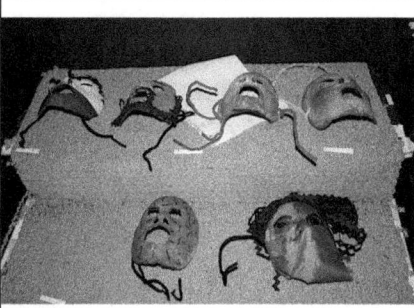

Is it made at the moment of activity, and (how) is it remade when it is shared, manipulated, cropped, edited for contrast, printed, hung and made to sing in a choir with other photographs as someone (or no one) looks on? If photography functions as an attempt at capturing, trapping, documenting, freezing time, is it as close as we can come to expanding and willing elasticity on time? Photography is urgent, and it is patient. Making an image – a photograph – forges *Kairos* despite and within *Chronos* time, pushing against seconds and minutes to create moments not bound by temporal shackles. (100)

... overdetermined about your attachment to the world?" How would you deal with the ways in which your normative fantasies about your attachment to the world actually don't describe all of the different ways you show up for it and maybe you could pay attention to those ways' (Berlant, 2016, 4:05–4:29)?

Embodied ways of knowing benefit from a multimodal orientation. Multiple practices of seeing and knowing are embodied. We hold our bodies in relation to other people, to the spaces we occupy, to the material conditions with which we interact. What are we 'trained to pay attention to'?

Multimodal ways of being (Vasudevan, 2011b) extend past the practices of bringing a multimodal lens to bear in the interpretation of texts or ...

... some of the other currently enrolled participants.

"People die. I know that's true."
How do you know?
"C'mon miss! Because they do!"
Yes, but how do you know that's true? What helps you understand something as true? Where do you place your trust?

Trey finally confided in Omar that his mother had "put him out the house", and left him to fend for himself. Trey had been staying with friends whose generosity was being tested as he was also a growing teenage boy with an appetite to match. He had left home with little, carrying his possessions in a small plastic bag, "the kind you get [at] a pharmacy", Omar described. He and Vicky reached into the emergency fund that Voices had to support exigent needs that arise – clean clothing and food were at the top of the ...

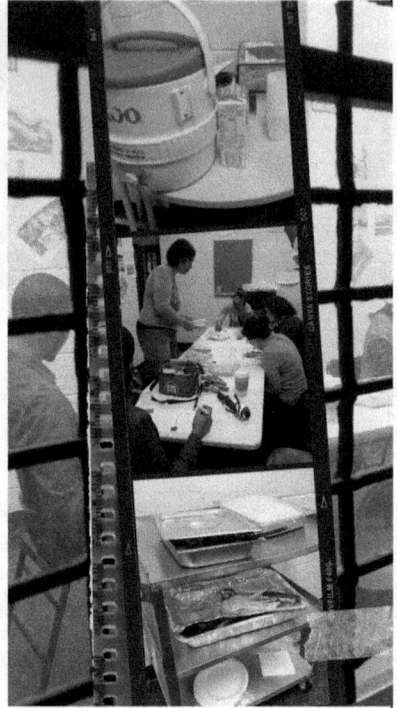

... collection of research data. Viewing the body as both canvas and instrument for composing meaning, a theory of multimodality can shape relationships among members of a research team, inform practices of attending to one another with hospitality, and focus attention on the affective resonances through which meaning is communicated. At the level of ephemera, there exist the daily pragmatics of embodying meaning such that to understand the story of a research partnership is to regard closely the ways that meaning is embodied and being enacted across the human, material and relational actors in research collaborations. The overt structures – schedules, policies, physical layout, curriculum – hold more than meets the eye.

Multimodal ethnography is full bodied listening. It's lying down next to a toddler to ...

... list – and he had purchased several items of clothing and some non-perishable food for Trey. He had made sure that all of the items were also placed neatly in a nicer bag, the one he had handed to Trey that day.
The bag.

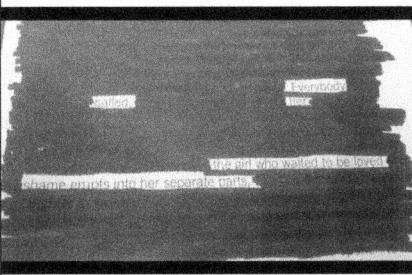

… sit on the carpet with her. It's observing what a teenager is conveying while remaining perfectly still; and recognizing the dollops of insight that are served up when the same teenager proffers ideas in prose, poetry, rhythm and rhyme. Is the body an artefact here? A text? A talisman?

'"Time" may appear to be an immutable construct, or one might even argue that we must naturalize our assumptions about time in order to move on with the real work concerning evidence and equity for learning'
(Leander et al, 2010, p 343).

Brite would fall asleep without fail every morning he attended our programme. Like clockwork. Sometimes, he would drift off to sleep with his head up and other times, most times, with the soft curls on his head nestled gently into the crook of his elbow.

The bag.

'Yet time, as it is lived and experienced socially and culturally, is constructed in specific and diverse ways; time is "made" rather than simply given in advance and filled up (Dubinskas, 1988). Time as "kairos," or our experience of time, is markedly different than time as "chronos," or clock and calendar time' (Leander et al, 2010, p 343).

Programme offerings were organized according to testing levels. The General Educational Development (GED) test operates as a high school equivalency exam. It is the centrifugal force that spawns pre-GED and post-GED courses. 'What comes next?' drives the agenda.

The bag.

	In an essay about his grandmother's shroud, author and photographer Teju Cole reflects on the embodied work of photographs at a time of sadness and remembrance.
	Photographs are there when people pass away. They serve as reservoirs of memory and as talismans for mourning (Cole, 2017).
	A photograph embodies –
	A collage embodies –
	A song or a playlist embodies –
	A soundtrack of laughter and jokes embodies –
	A relationship, a moment of joy or pain, discovery, a new ritual, friendship, rivalry, a moment of realization/learning/recognition.
	'A photograph insists on raw fact and confronts us with what we were perhaps avoiding' (Cole, 2017, para 7). (100)
'And yet clock and calendar time – whether truncated at the event or extended and divided up into units at greater "scale" – is what orders and structures research and guides our notions of "development"' (Leander et al, 2010, p 343).	Stories of how projects were founded, lived and reformed are also embodied … while writing a research blog collaboratively. … while making food together for the ritual of family night with participants and their families. … while listening to someone talk and attending to their hands, moving and still. … while watching a young person take a photo and find delight in what they have made. … while re/forming research questions to respond to unfolding dynamic shifts.
	While … a temporal space we inhabited as mutually vulnerable researchers *with* participants in the pursuit of making knowledge while remaking knowledge spaces. (100)

After-word

Writing is seeking.

Seeking time, flow, images, metaphor, memory ... to craft just the right words – or, sometimes, and more often now than before, just the right images and movements – to cause change in perception, to craft a sense or feeling in the hopes that the reader is transmuted into a narrative collaborator.

The unpredictable reader, the unreliable reader, the unknown reader.

While working through another revision of this chapter, I heard the news that Lauren Berlant had passed away. I only arrived at her work a few years ago, in part through happenstance and in part through a recommendation from a colleague. Her age at the time of death, 67, gnawed at me. What else might she have written? What other questions might she have posed? What else ... These are the questions we ask of those who have moved us, and in the imagining and eventual writing of this book, many of those who have pushed, inspired, and transformed our thinking have passed away. Some were teachers, others collaborators, some more distant than others, but all were mentors of a type. Each time, I took solace in Maya Angelou's poem 'When Great Trees Fall' (1991)[1] that concludes with this stanza:

> And when great souls die,
> after a period peace blooms,
> slowly and always
> irregularly. Spaces fill
> with a kind of
> soothing electric vibration.
> Our senses, restored, never
> to be the same, whisper to us.
> They existed. They existed.
> We can be. Be and be
> better. For they existed.

Berlant's questions (shared previously) of where, to what and how we attend – pay attention – is at the heart of how research unfolds. What matters? And how is that determined? Far from being rhetorical or existential questions alone – for there are real answers for the traditions of research in which we engage – Berlant's musings call into question tropes and practices that appear to be adopted unquestioningly.

---

[1] Maya Angelou wrote the poem, 'When Great Trees Fall', in 1987 on the occasion of James Baldwin's death and read it at his funeral. It appears in Angelou's volume *I Shall Not Be Moved*, and was originally titled, "Ailey, Baldwin, Floyd, Killens, and Mayfield."

Another few clicks of the keyboard reminded me that *The Hundreds* was one of the first texts Richard, Kate and I (re)visited together when we began writing this book. In *The Hundreds*, Berlant and her co-author, Kathleen Stewart, craft 'exercises in following out the impact of things (words, thoughts, people, objects, ideas, worlds) in hundred-word units or units of hundred multiples' (Berlant and Stewart, 2019, p ix). The hypnotic repetition of form[2] produces additional layers of meaning, paying homage to and pushing the boundaries of discursive pastiche and astute commentary; and, in a way, resonating with hyperlocal inclinations of ethnography (Coffey, 1999). In doing so, Berlant and Stewart allow form to subvert the normative practices of consuming a text.

W.G. Sebald, at a reading of his work[3] at the 92nd Street Y, a cultural and community centre in New York City, touched on a similar vein about form when he responded to a question about his trademark practice of including photographs in his books:

> [Photographs] hold up the flow of the discourse. As one knows, as a reader one tends to go down this negative gradient with a book that one reads towards the end, so books have almost by definition an apocalyptic structure. And it is as well [the function of photography] therefore to put weirs[4] in here and there [that] hold up the inevitable calamity. (15 October 2001)

Likewise, the form that research takes, as it is often described in methods books, is stretched, tested, bent and pushed by interlocutors who are guided

---

[2] Moved by Berlant and Stewart's example of 'hundred-word units', I revisited one section of this chapter (the top right quadrant on the first five pages of this chapter) with a desire to bring the expanse of photography into focus within a discussion about how images can embody the spaces in which they were made. Similar to Kate and Steve's reflections in a recent article, I sought to craft 'fragments of writing' that 'are not findings, or examples; rather, they should be taken as they were intended, as small stories or "telling tales"' (Pahl and Pool, 2021, p 662).

[3] In October 2001, the author W.G. Sebald gave a reading of his book *Austerlitz* (2001), which had just been translated and published in the US (from the original German). Sebald's writing has been described as genre-less or, in other contexts, genre-bending, as he blends historical facts and narrative together with fictionalized portrayals of people and events. His writing does not adhere to the familiar conventions of history, fiction, historical fiction, memoir, or travelogue, and often blends all of these approaches together to yield a uniquely Sebaldian textual experience. His use of photographs, some of which he finds in history archives and others that he produces himself, sometimes includes various forms of manipulation and editing. I have found his writing to be ethnographic, lyrical, arresting and poetic.

[4] A weir is a dam built across a stream or river to raise the water level or divert its flow.

by a desire to co-produce knowledge. In our project, ethnographic methods of participant observation and interviews were infused with arts-based methods of creating artefacts – collaging, impromptu video making, shared text-making, photography; these practices, as they become normalized within our research ethos, served as 'weirs' and interrupted the flow of received meanings that are embedded often in the labels ascribed to young people by the systems of schooling and justice.[5]

While attending to the ways that research projects can serve as sites for the emergence of moments and artefacts that both show and tell – or as sites for what Kate Pahl and Steve Pool (2021) discuss as 'research-creation' – as researchers, we are imbued with the promise of writing and *righting* stories heretofore scripted within the confines of institutional labels. *The Hundreds*, therefore, could also be read as a different story of constraints, of being exact and exacting, of showing and telling something about a moment, a person, a consequence of how observations about life unfold amidst systemic and passing forces.

Rainer Maria Rilke's (1993) advice to a young officer in *Letters to a Young Poet* holds respite from systemic constraints, not an escape but rather as a reimagination:

> Everything is gestation and then bringing forth. To let each impression and each germ of a feeling come to completion wholly in itself, in the dark, in the inexpressible, the unconscious, beyond the reach of one's own intelligence, and await with deep humility and patience the birth-hour of a new clarity: that alone is living the artist's life: in understanding as in creating. (p 23)

If read solely as prescription for an 'artist's life', narrowly defined, the implied meaning – to live artfully as humans – would be lost. Collaboration between research partners committed to co-production demands a capacious artistic sensibility that can view relationships, too, as 'gestation and then bringing forth', without which such work – especially, participatory ethnographic research – is not possible. Research of this type, as a practice of witnessing with – or, with-ness (Vasudevan, 2016; Rodriguez Kerr et al, 2020) – is

---

[5] In their now-classic text, *Successful Failure: The School America Builds*, Hervé Varenne and Ray McDermott (1998) critique the practices of labelling and categorization that – at the time and is still true today – dominate American schooling. Categories and labels are culturally produced at particular moments and given particular circumstances, but when they become embedded in an ongoing way into the practices of an institution it becomes difficult for children to avoid being consumed by them. The authors dive deeply into the cultural production of 'disability' as a category assigned to a child whose practices did not conform to school expectations. Given the 'onslaught of special education designations for children in American schools', that child, 'Adam', was fated 'to be acquired by one of them' (p 27).

rooted in mutual vulnerability. The impetus to pursue forms that embody – represent, convey, evoke – is sown in that vulnerability.

If writing is seeking, then writing about something or someone – an experience, a project, acquaintances, collaborators – is the act of seeking something approximating understanding through representation. Is that what we do when we seek (the appropriate forms to embody and through which) to represent our research? This chapter was an attempt to explore that question.

'Findings' are an inadequate subheading for what emerges from work that renders researchers and participants mutually vulnerable as is the case in the collaborative forms of research in which my colleagues and I were engaged. Still, we use this and other terms – conclusions, lessons, emergent themes – to offer legibility to the point at which we arrive after weaving stories from our project together with the theoretical wanderings of others (in other words, analysis). That point is messy, frayed, sometimes fractured and fractalized, but it holds something that speaks to the temporal moment of its conception. Sometimes, the end point of a project is forced by the conclusion of a grant cycle or the completion of funding for a thesis. Sometimes, this moment is brought on by changes at the organization. Sometimes, life intervenes and a project is paused, indefinitely. And even as a project comes to an end – that is to say, that the ways of being and doing that were associated with a group of people and a site cease to exist in the same way as before – the legacy remains embodied in the texts, artefacts, and private and public memories that were a part of the experience and those that follow.

Findings and words like it are an attempt to orient and prepare the reader to receive something that the writer is conveying with painstaking care that aims to hearken back to themes while looking ahead to implications. Findings reveal the inadequacy of words, alone, to communicate the journeys of enquiry. In collaborative and participatory research spaces, where words are overlapping and voices interject emotion into specific places and times, capturing the 'what' of an event – the shutter release at the exact right moment (that is, the moment of the desired action); or jotting down some of the words, some of the spatial elements, some of the materiality – is simultaneously a futile-seeming and ultimately necessary pursuit. Virtual and augmented reality simulations, 360° cameras, wearable technologies and recording devices make the practice of capturing 'what is going on' ever more possible, in a sense. But one wonders whether the ability to create more versions of documentation results in clearer or just more fractured pictures of how a space unfolds. Embodiment is a word that tries to capture what is left of a project. This chapter, even with the inclusion of words, images and a cacophony of form, still offers only a partial glimpse of the affective traces that took root in the circle within us and the one that we draw around us, together.

INTERLUDE 6

'Most people don't believe me'

Katie Scott Newhouse

What is the materiality of believing?

Is there such a thing?

A way that material objects offer the same validation as another's belief.

Who created the Belgian blocks that Lalitha is standing on when you tell her 'The first time I had a gun in my hands was when I was arrested.'

She didn't respond immediately. Was it that she did not know how to respond?

Or was it that she did believe you?

And could not express in words the weight of her belief,

Heavy like the Belgian blocks which fill the street where you both stood.

Over 100 years ago, a person, probably more than one, laboured to cut and lay these stones, often confused with cobblestones but given their own name: Belgian blocks.

Intentionally designed for the width of a horseshoe.

Here the stones remain. The people who laid these stones believed things.

I wonder who else has asked, 'do you believe me?' in this same spot.[1]

What are the things that people do and don't believe about you?

---

[1] I am drawing on Katherine McKittrick's (2006) book *Demonic Grounds* which discusses how space, time and representation take on different meanings, for a variety of different reasons, often connected to an individual's lived experience. There is an embodied quality to standing on a 100-year-old street while having a conversation with a young person that requires vulnerability from both the adult and the youth. This interaction may occur in and around the same geo-physical location as where the Belgian blocks were

How often do you linger on the need for that belief?

As if someone else's understanding validates your own experience.

In many ways it does. When others believe us, our realities become clearer, more tangible.

Through the validation of belief we are now part of a shared constructed spatial reality.

A space that is more than a physical location.

Space without place.

'Your ancestors killed mine,' Roger tells me.

In this moment *I* hesitated,

to express that I deeply believed his statement.[2]

When words fail, what else remains –

The spaces?

Weeks later, Roger told me:

'I'm changing, *slowly*'

Sitting across from one another

in a converted office space housed in a formerly historic hotel.

Like the Belgian blocks, bricks of the former hotel were laid over 100 years ago.

Roger's words linger in the air between us,

In my mind colliding with other stories from young people.

About adults, not believing.

Telling young people to give up and drop out of school.

Roger saying, 'I'm changing slowly, *because* I finally found someone to listen to me.'

Does listening pre-suppose belief?

or is it the other way around?

The material conditions for believing. What is needed for belief to thrive?

In the vulnerability of the question, '*do you believe me?*'. And the qualifying statement, '*most people don't*'.

What is the materiality of believing? Is there such a thing?

Carving out spaces where youth and adults are vulnerable together.

Creating spaces which serve as containers for this act of believing.[3]

---

laid 100 years ago, but meaning is made as a person intersects with their own historical/ contextual understanding *and* the material artefact.

[2] I discuss this experience more in-depth in a piece I co-wrote with Kristine Rodriguez Kerr and Lalitha in 2020 about conducting participatory multimodal ethnography with youth.

[3] I wrote this poem in response to Lalitha's writings as a found poem where I weave together the words she shared with us and a few of my own thoughts and wonderings. I focused specifically on this idea of belief and what it means when a young person asks an older person, "do you believe me?". This poem serves as a textual reflection of the ethos we

The physical components of a space:
The Belgian blocks,
Former historic hotels
Physical materials used to create spaces
to hold moments for
belief in one another
together.[4]

---

created as a research group, built upon the idea that kids matter, and what kids do and say also matters (Rodriguez Kerr et al, 2020).

[4] This remains a central tenet for my own research practice as an early career scholar engaging with participatory multimodal frameworks along with Disability Studies in Education (Gabel, 2005) and Critical Spatial Theory (Massey, 2005; Soja, 2010). Crafting this poem as a response to Lalitha's vignette about Robert provided me with a moment to consider my own lived experiences as a graduate student researcher and workshop facilitator. This poem serves as a tracing of how participating in the 'Reimagining Futures' (RF) project helped me to crystallize my own commitments to conducting research with youth within educational spaces and how to engage in that work from a person-centred, youth-valuing and arts-based way. As I reflect on my experiences as an RF team member and now fully credentialled doctor of education, I consider how the collaborative research of the RF team influences the spaces I am creating with colleagues and graduate students to explore the complications of conducting multimodal scholarship and research with young people. In my own research, I continue to wonder about and explore the materiality (sometimes virtually during the pandemic year) of gatherings with both youth and adult collaborators and co-conspirators.

7

Hypertext

Richard Steadman-Jones

Kate:	What other modes of writing have you used to capture these projects?
Richard:	Well, one of them is hypertext, which is a form of writing that can capture the multiple voices that speak within the projects.

Hypertext is a species of text defined by its structure. A hypertext consists of an array of textual 'chunks' – following the usage of Roland Barthes they are often called 'lexias' – and these lexias are linked together in a more or less complex way. Typically a hypertext will make use of the digital medium. It certainly is possible to produce a printed text with a hypertextual structure but some critics reserve the term 'proto-hypertext' for such works. Generally speaking, one reads a hypertext on a screen and navigates through it using a system of hyperlinks. The following diagram provides an example of the way in which a set of lexias may be joined by links:

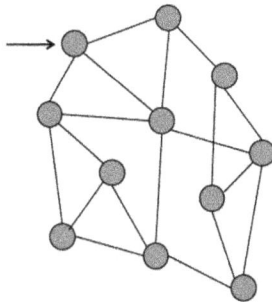

Hypertext structures offer the reader the possibility of choice. Reading a printed novel is usually a linear experience. One starts on page 1, reads page 2, and so on until the end. However, the opening page of a hypertext will typically give the reader several options, as will each of the lexias subsequently encountered. Thus, the text offers the reader multiple pathways and the particular pathway taken in any act of reading is entirely up to the reader.

Early in the history of hypertext criticism, there was some excitement at the idea that the hypertext surrenders control to readers. The linear structure of the printed novel was seen as one in which the author remained firmly in control of the unfolding narrative, whereas the hypertext with its multiple pathways was seen as prioritizing the decision-making of readers, placing choice at the centre of the experience. In practice, hypertexts vary in the extent to which their links are transparent and hence the extent to which the reader's decisions about which pathway to take are made with full information. Links may straightforwardly state the destination to which they point, in which case the reader's decisions about where to go are well informed and grounded. However, the links need not include so much information and this will produce a more arbitrary experience of reading – readers may feel that the pathways they take are to a large extent a matter of chance.

We began to work with hypertext during the 'Communicating Wisdom' project – the one that focused on fishing. A feature of this project was the multiplicity of voices that spoke within it. The anglers themselves talked to us about their experiences. The men taught the young people the essentials of the craft: tying the hook to the line, casting, reeling in, releasing the fish back into the water. The young people turned up most weeks, gathered by the pond, and sometimes talked to us: "We sat under an umbrella", said one. "It were pouring down. We didn't care, did we? … We wanted to catch a fish." Looking at these words now – years later – I notice the question, "We didn't care, *did we*?". The speaker is not telling the story alone but turns to those around her for confirmation of the strength of their passion. It is a piece of co-narration about the power of desire. "It were pouring with rain", she continues, "That's another world, the pitter patter and pouring off the edges and then you see it from a different angle". The emphasis on change is important here. The unaccustomed experience of sitting out in the rain, partially protected by a large umbrella, triggers a shift in perception. You see things in a different way and this is what fishing does – it changes how you experience the world.

However, the utterances of the young people were not the only ones that resonated within the work. The team included a philosopher who was particularly concerned with the concept of wisdom and the idea that the transformations wrought by the practice of fishing might be seen in terms of a process of 'becoming wise'.[1] Another participant was a poet who assembled a collection of contemporary texts, all of which focused on fishing, and wrote about the ways in which each engaged with a particular concept of sublimity. Still another wrote a daily blog about the mythology of fishing, which included stories from Ireland, Iran, Japan and Scandinavia. And at the same time, we were all reading Izaak Walton and Charles Cotton's (1676/ 2008) poetic, contemplative, didactic work *The Compleat Angler*, which was first published in 1653 and which is simultaneously a 'how to' book about angling and an argument for the pleasures of the contemplative life. As the project progressed, we began to sense connections between all of these different discourses. The 'seeing from a different angle' described by the young speaker on the edge of the pond resonated with the philosophical, poetic, mythological voices articulated by other participants and indeed with Walton and Cotton's famous text.

In Chapter 2, we introduced the concept of polyphony and this particular project seemed to have a strongly polyphonic character. The different voices entered into what Mikhail Bakhtin called a relationship of dialogism and the work of the project was not simply to process the utterances of the young people – classifying them and coding them with a tool such as NVivo – but to bring them into an interaction with other kinds of discourse. Hypertext constituted a way of staging this interaction, the individual lexias consisting of sections of text taken from all of the different streams of work. Some quoted the young people. Some were taken from ethnographic fieldnotes written after sessions on the edge of the pond. Some presented philosophical ideas. Others were excerpted from the mythological blog and still others included images. Each lexia of the hypertext is linked to five others but the links are not transparent – they do not constitute pieces of text that tell you unambiguously where the link will lead. They are abstract signs – all identical – and so when readers decide where to go next, they do so on an arbitrary basis. The result is a labyrinthine text. One wanders through it constantly making decisions about where to go and never knowing what will greet one when one arrives. After reading the words of the young people, one can find oneself confronted with an account of Ernst Bloch's theory of wisdom or a story about the Irish hero Finn McCool.

As one wanders, connections emerge between the textual fragments. The young people seem to be speaking of philosophy. The mythological stories touch upon the same emotions as the fieldnotes. Despite its fragmentary character, the hypertext emerges as a singular text, the individual lexias entering into relationships as the reader travels through them. To some extent, this is an effect of the author's making. Many of the links are carefully

---

[1] Specifically, his work focused on Ernst Bloch's conception of Wisdom (see Bloch, 1969).

considered and are intended to bring particular pairs of lexias together. However, it is not really possible for the author of a hypertext to anticipate all the possible pathways that readers will take though it. The mathematics of the work militates against it: if the first lexia opens on to five others and each of those onto five more, the number of possible paths increases exponentially with each click of the mouse. Hence there is a kind of serendipity to the functioning of the hypertext. It constitutes an invitation to read together a number of different threads of work – it calls upon the reader to investigate the dialogic character of the material.

Following are three collections of lexias. Each represents a pathway that a reader might take through the hypertextual work. Which one readers encounter depends upon the choices they make. All pathways begin with the same initial text – a quotation from the opening of *The Compleat Angler*. This quotation points to the connection between fishing and wisdom, as well as the notion that fishing involves the contemplation of nature.

All pathways begin here

An ingenious Spaniard sayes, 'That both Rivers, and the inhabitants of the watery Element, were created for wise men to contemplate, and fools to pass by without consideration.' And though I am too wise to rank myself in the first number, yet give me leave to free myself from the last, by offering to thee a short contemplation, first of Rivers, and then of Fish: concerning which, I doubt not but to relate to you many things very considerable. (Walton and Cotton, 1676/2008, pp 40–41)

Pathway 1

I

In his essay '*Über den Begriff Weisheit*', Bloch identifies the following dimensions as belonging to wisdom:

- Becoming peaceful. Wisdom is not cleverness, but a fundamental calmness that must not be confused with inactivity. Not-knowing.
- Maturity. 'So preciously distinct from common sense.'
- Active care for the common good. Wisdom is partisan.
- Imperturbability. Responsible for the caricature of the sage. Bloch re-reads it as health.
- Tao or simplicity. The effortless act, in tune with the tendency in things. Tao means way. Wisdom is streetwise.
- Beyond. In traditional wisdom, a link to the un- or other-wordly. For Bloch, the prophetic dimension of wisdom: transgressive speech;

imagining the alternative; conformity of the will with a final, but open, purpose.

- Humour. A mode of truth-telling in itself. Its lightness is the utopian side of wisdom. (Johan Siebers, Notes on Ernst Bloch's Conception of Wisdom, June 2014)

II

I sat with Reece and he began to tell me lots of interesting things about fishing and giving me tips about what to do by talking through what he was doing. He told me about the ghost carp and he said that he would love to catch one because they are white and rare. He told me about barbel, pike, and koi carp. He said that barbel have whisker-like things called 'barbules' around their mouths. Pike are predators and eat everything and there is one pond that was paying fishermen to catch the pike because they were eating all their stock. He said that koi carp were beautiful. (Hugh Escott, fieldnotes, June 2013)

III

In *Aion* (which has the subtitle, 'researches into the phenomenology of the self'), Carl Jung has much to say about fish as symbols and a little too about the activity of fishing. He talks about the 'primitive totemistic identity between hunter and prey', citing the examples of 'the Babylonian culture-hero Oannes', who 'was himself a fish', and the 'Christian Ichthys', who is 'a fisher of men par excellence'. Then, complicating the picture, he says of Christ-Ichthys, 'Symbologically, he is actually the hook or bait on God's fishing-rod with which the Leviathan – death or the devil – is caught'.[2] (Richard Steadman-Jones, fishing blog, 21 February 2013)

IV

[T]he great Naturalist Pliny sayes, '[That Natures great and wonderful power is more demonstrated in the Sea, then on the Land.]' And this may appear by the numerous and various Creatures, inhabiting both in and about that Element: as to the Readers of Gesner, Randelitius, Pliny, Aristotle, and others is demonstrated: But I will sweeten this

---

[2] The quoted text is from C.J. Jung (1968) *Aion: Researches into the Phenomenology of the Self*, translated by R.F.C. Hull, 2nd edn, Princeton, NJ: Princeton University Press, p 174.

discourse also out of a contemplation in Divine Dubartas, who sayes [in the fifth day],

> God quickened in the Sea and in the Rivers,
> So many fishes of so many features,
> That in the waters we may see all Creatures;
> Even all that on the earth is to be found,
> As if the world were in deep waters drownd. (Walton and Cotton, 1676/2008, p 44)

V

Many koi carp are orange and they are expensive. I asked whether they were ornamental and whether he had caught one. He said that he knew ponds where you could catch them. We talked about the myth that I half remembered about koi carp swimming up a waterfall and turning into dragons, and he said that he had heard of it. I joked that you wouldn't have a myth like that about a pike. Reece told me that there were breeding places where you could pay a tenner and go and look at all the fish in the tanks. (Hugh Escott, fieldnotes, June 2013)

VI Tattoo

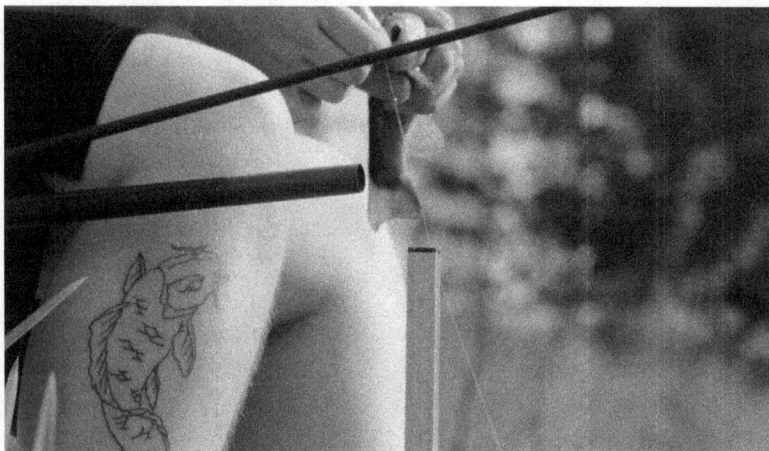

Source: Photograph by Steve Pool

VII

I felt like the evening went on for a long time. Jean got hungry for her tea. Steve, Dylan, and the others sat still and caught fish while I watched moorhens and gulls. The water plopped with flies and fish and glittered in the evening sun. The fishermen drank the tea they had brought in a thermos. Sometimes we got cold and put on our jackets but in the sun it was lovely. It was as if nobody could drag themselves away from the bank. Dylan was in an amazing rhythm of catching fish. Martin smiled and told me of his early morning fish that day. Jean told me about her worries about her job. I told her I wanted to be a river board man when I grew up. We became a group of figures in a landscape. (Kate Pahl, fieldnotes, June 2013)

Pathway 2

I

Here are no entrapping baits,
To hasten too, too hasty fates,
Unless it be
The fond credulity
Of silly fish, which (worldling like) still look
Upon the bait, but never on the hook. (Walton and Cotton, 1676/2008, quoting Henry Wotton, p 225)

II

There is a Buddhist text called 'The Fisherman' – 'Balisika Sutta'. It is part of a collection of scriptures that originated early in the history of Buddhism and is written in Pali, a South Asian language close to that which the Buddha himself spoke. The sutta begins with an analogy: '[J]ust as if a fisherman were to cast a baited hook into a deep lake and a fish with its eye out for food would swallow it – so that the fish that had thus swallowed the fisherman's hook would fall into misfortune and disaster, and the fisherman could do with it as he will – in the same way, there are these six hooks in the world for the misfortune of beings, for the slaughter of those that breathe.'[3]

The six hooks turn out to be associated with the senses as they are conceptualized by the tradition: the five that we are familiar with – sight, hearing, touch, taste, and smell – as well as the intellect, the sixth 'sense' of Buddhist psychology. (Richard Steadman-Jones, fishing blog, 22 February 2013)

---

[3]   This translation of the Balisika Sutta is by Thanissaro Bhikku: www.accesstoinsight.org/tipitaka/sn/sn35/sn35.189.than.html.

III

My epiphany is that the insight we are searching for is wrapped up in the stories we tell and that the meaning which we are making is not about reducing these stories to fieldnotes or films or seminars but in collectively finding a way to layer and fold the stories which tell a slanted truth about the world. (Steve Pool, email, 10 July 2013)

IV

Fun: Mythologies develop naturally around fishing. The carp that saw right through the trick of the bait. Even Reece's tale of 'breeding places where you can pay a tenner and look at all the fish' carries an aura of enchantment and exotic mystery, a far-away place where money will really buy you happiness. Where no one guards their happiness so carefully that it cannot be expended in exchange for some cash. Enough to go around. Who knows if such a place really exists? It is a place of breeding, of origination. The fountain source of Alph at the heart of the pleasure dome.

There is a curious dimension to wisdom that has to do with imagination. It seems that a part of what we call wisdom can only be accessed by the imagination, not by concepts, not by praxis, not by prudence. There is a creative dimension to wisdom, it opens up onto the not-yet, onto what might be and what it might be like.

Wisdom's mood is the subjunctive, the reverie. (Johan Siebers, Notes on Ernst Bloch's Conception of Wisdom, June 2014)

V

I went over and talked to the fishermen. They told me stories. One told me about working in a high-level internet training firm and how the way they did things was different from in a classroom – there were just these amazing experts on hand to help. This is how fishing is taught.

Martin said he had got up at 4:30 that morning and he had gone fishing. He said Steve would have liked it as the mist rose in the water. (Kate Pahl, fieldnotes, June 2013)

VI

There is a paradoxical quality to the image of Christ-as-bait. Passiontide is a dark time in the Christian calendar but, still, the death of Christ is seen as needing to take place in order to bring about redemption. And so, in Jung's interpretation, God is characterized as offering Jesus as bait or as a lure to death and the devil so that – when they bite – they will be defeated. In a footnote to the text, he cites Saint Cyprian, the third-century bishop of Carthage: 'Like a fish which darts at a baited hook, and not only does not lay hold of the bait along with it, but is itself hauled out of the sea; so he who had the power of death did indeed snatch away the body of Jesus unto death, but did not observe that the hook of the Godhead was concealed therein, until he had devoured it; and thereupon remained fixed thereto.'[4] (Richard Steadman-Jones, fishing blog, 21 February 2013)

VII

I have got sneaky with the fish because they are quite sneaky and hidden. (Angler, Rotherham, June 2013)

Pathway 3

I Hands

Source: Photograph by Steve Pool

---

[4] The quoted text is from C.J. Jung (1968) *Aion: Researches into the Phenomenology of the Self*, translated by R.F.C. Hull, 2nd edn, Princeton, NJ: Princeton University Press, p 174, n 38.

II

Angler: Round the goalposts, push him down, wet him, like Terry says. You wet him because if it's dry and you pull nylon, it will burn, the friction will burn it. Pull it slowly.

Young Angler: It's right hard that.

Angler: Keep going there until you see. Pull down tight, right? And release it. Pull him through and there he is.

Young Angler: Right.

Angler: Think you can do that? (Anglers, Rotherham, June 2013)

III

Socrates was killed by his community.

Philosophy, literally: the love of wisdom. Yet this word, 'wisdom', has almost been eradicated from philosophy, as it has also been eradicated from our culture in a more general sense. We value skills, we value knowledge, we value information, sometimes we value faith. But wisdom is something else than all of these, including the valuing itself, although they can all be part of it. Wisdom has something to do with the ability to integrate the intellect and the will in the experiment of life.

Mikhail Epstein says that wisdom 'mediates between the virtues of patience and courage, helps us to accept what we cannot change, and to change what we cannot accept'. Wisdom includes an awareness of its limits and its humbleness, otherwise it becomes foolishness: Socrates was the wisest of all because he knew his wisdom was worth nothing. Erasmus wrote in praise of folly. There is foolish wisdom and wise foolishness. (Johan Siebers, Notes on Ernst Bloch's Conception of Wisdom, June 2014)

IV

Terry told a story about a chap who was fishing from a bridge smoking a cigarette and he fell from the bridge into the water, thigh deep, and he continued to smoke his cigarette, undaunted. (Hugh Escott, fieldnotes, June 2013)

V

Men that are taken to be grave, because Nature hath made them of a sour complexion – money-getting men, men that spend all their time, first in getting, and next in anxious care to keep it – men that are condemned to be rich, and then always busy or discontented: for those poor rich men, we anglers pity them perfectly, and stand in no need to borrow their thoughts to think ourselves so happy. (Walton and Cotton, 1676/2008, p 22)

VI

I fished very badly today. I kept losing the bait. Fish were popping all around me but they would not bite. Jean said, "Oh, they are wicked". She screamed when she saw a water snake and would not come near the bank. I stayed but felt useless and also was convinced I did not have what it took to catch a fish. (Kate Pahl, fieldnotes, June 2013)

VII

Here is the speech by which God reminds his suffering servant of the extent of his power in the book of Job 41: 1–10:

'Canst thou draw out leviathan with an hook? Or his tongue with a cord which thou lettest down?

Canst thou put an hook into his nose? Or bore his jaw through with a thorn?

Will he make many supplications unto thee? Will he speak soft words unto thee?

Will he make a covenant with thee? Wilt thou take him for a servant for ever?

Wilt thou play with him as with a bird? O wilt thou bind him for thy maidens?

Shall the companions make a banquet of him? Shall they part him among the merchants?

Canst thou fill his skin with barbed irons? Or his head with fish spears?

Lay thine hand upon him, remember the battle, do no more.

Behold, the hope of him is in vain: shall not one be cast down even at the sight of him?

None is so fierce that dare stir him up: who then is able to stand before me?' (Richard Steadman-Jones, fishing blog, 12 March 2013)

★★★

Running across the lexias is an exploration of the connections between wisdom and the practice of fishing. There is the way in which the fisherman 'becomes fish'. To catch a fish, you must think like a fish – you must 'get sneaky' because the fish are sneaky – and, indeed, Christ is both fish (because the initial letters of his title *Iesus CHristos (H)Uios THeou Soter* spell the Greek word for fish, *Ichthus*) and a 'fisher of men'. There is the elusive nature of the fisherman's craft – the anxiety that one does not have 'what it takes to catch a fish' and God's reproach to Job, who does not have the power to catch the wild creature that is Leviathan. There is the hook and bait as symbols – their appeal to 'silly fish' an image of the sensual lures that tempt humans in the chaos of the world. There are the qualities of the fisherman – calm, quiet, imaginative and receptive to the pleasures of the simple life. These themes run between the lexias emerging little by little as the reader navigates the hypertext.

At root, the hypertext defamiliarizes the textual fragments that make it up. When they appear as part of a varied constellation of material, they strike the reader differently from when they appear contextualized within their original sites. Navigating through them constitutes a kind of provocation. It calls upon the reader to find the connections between and across very differently constituted elements. The subtitle of this book is 'the poetics of letting go' and a hypertext is, in a sense, a type of 'letting go'. To return to a point made earlier, the author of a hypertext is not always in complete control of the reader's progress through it simply because – if the text is densely linked – after a few clicks, the number of pathways becomes too great for anyone to anticipate. Hence, there is an unpredictable quality to the connections which emerge. The text operates like a mechanism for divination – for practices which Maggie MacLure (2021) describes as 'diagrammatic, ambulant, cryptic, and experimental' (p 510). It has something of the quality of the I Ching, which was – of course – of central interest to that pioneer of unplanning, John Cage. And it works to re-enchant material bringing to it a quality of vibrancy and poetry. The hypertext stages the creative encounter between voices and these voices acquire new meaning through their interaction. It is a polyphonic mode of writing which reflects the multiple vocality of the work itself.

INTERLUDE 7

Failing

Andrew McMillan, Kate Pahl and Vicky Ward

Following are four 'critical incidents' which tell the story of a project that was called 'Taking Yourself Seriously' (see Appendix). Its aim was to explore the impact of the work of a group of artists (poets, musicians, visual artists) within a secondary school. The project team aimed to co-produce with a group of young people from the school a research investigation on the impact of this work. The project took place over a year, directly after the Brexit vote in the UK, and within a school that had experienced some racial tensions. It had also been characterized as a 'failing school' by the UK inspection regime, the Office for Standards in Education (OFSTED). The school was situated in a small town in the North of England. The town where the project was situated had itself experienced hardship after the closure of the mines in the 1980s, including a decline in jobs, and was seen as an area that was 'left behind'.

We tell these stories drawing on small-scale descriptions of our experience in the school, which we have called 'critical incidents'. Following Kathleen Stewart (2007), we consider the 'ordinary affect in the textured, roughened surface of the everyday' (p 39). This brings the everyday into the writing as well as acknowledges our experience of the project. We begin with the following incident as, for us, it provides an introduction to the school and locates it, nested within a community that was struggling to accommodate a number of fractures and divisions.

Our project involved a research team of six young people, who were supported by Vicky Ward, a social worker and worker on this project. The young people met on a Friday morning and they were tasked with documenting the impact of the artists on them. They were given the choice of documentation process, and they produced, with images and words, an account of this project in a jointly produced booklet. In this account, their names are anonymized.

Critical incident 1: 'Ordinary affects' by Vicky

Kate picked me up from Sheffield train station early one cold, crisp October morning so that we could get there before 9am in order to be ready to facilitate the workshops with the research group. As we drove down the final road to the school, an older man staggered out and struggled to reach the bollard in the middle of the road. Having worked in substance misuse services for several years, instinct told me something wasn't right. Kate and I had a very brief conversation as I was getting out of the car, and I agreed to meet Kate at the school. The man took my arm and slumped to the ground as we made it to the pavement. A kind taxi driver, a local man from the Muslim community, stopped and offered to call an ambulance, and the woman from the florist popped over with her body warmer.

The man was withdrawing from alcohol and had gone out without taking other much-needed medication. The emergency phone-line operator said the ambulance would be there as quickly as possible. An hour later, we were all still sitting on the pavement. The man in withdrawal had been to the shop to buy some alcohol so was already stocked with today's supply. Another hour passed and we're still sitting on the pavement. The man on the floor says that he's starting to feel a bit better, but is very cold. The taxi driver offers to take the man back to the sheltered accommodation where he lives. We rang them to try and get hold of the warden but no response. The taxi driver refused to take him without me coming along to help. I explained I needed to get to school as soon as possible as I was supposed to be teaching. Other passers-by stopped and offered to call for an ambulance; they were shocked when we said we'd already been waiting nearly two hours.

The man on the floor said he thought he could make it back to his place in the taxi. I reluctantly agreed to go with the taxi driver. He also promised to drop me back to the school straight afterwards. We phoned the ambulance and cancelled the call. The lady in the florist came back over as she wanted her body warmer back. It belonged to her late mother so she wasn't prepared to let someone else have it just yet.

We arrived at the sheltered accommodation and couldn't find a warden anywhere. The man found his front door key and we helped him back into his home. After making sure he had access to his medication, a phone, and knew where the emergency alarm button was, we left. The taxi driver dropped me back at the school and we all went back to work.

The whole event was frustrating but somehow emblematic of our experience of the project in the school. Instead of sitting down talking with young people about cohesion, I find myself living cohesion: negotiating what could have been complex interactions within a community fraught with tensions around issues of race and class and gender. Instead, we all worked together to make sure a vulnerable member of the community was kept safe.

Reflection

Our response to this is to think in moments, in feelings. These moments were emergent presents that felt like something. We experienced our project in moments, and bodily. When a young person got up and taught poetry to a class, his 'stance' was that of a teacher. Affectively, we lived the project through these moments. Our experience of going to the school was affected by our early starts, our conversations in the car on the way there, the experience of waiting in the lobby and constantly having to persuade the staff that we knew what room we were in, and our many mistakes – over-ordering equipment, managing not to produce lesson plans, and our loss of key objects, such as a diary, all contributed to the affective experience of the project. The ordinary did not stay at the door but came in with us. This then has implications for pedagogy as Christian Ehret (2018) argued:

> Thinking-feeling moments as pedagogical orientation require being in the moment with youth, and imagining together what youth can do and who they are becoming. Being in moments with youth therefore challenges a representational image of pedagogy, which is composed of linear progress and narrow outcomes that constrict the purpose of education and reify inequities. (p 56)

Drawing on this idea of non-linear and affective experience, we introduce our second critical incident, by Andrew, about creating in a school.

Critical incident 2: 'Creating in a school' by Andrew

One of the things about any kind of creative activity is that it is, by its nature, oftentimes loud and chaotic, and this puts the space of creativity into conflict with the rules and regulations of the 'failing school' attempting to enforce discipline; the overwhelming quietness and calm of the corridors was one of the first things I noticed when entering the school. Creative practice as a mode of engagement with young people can be complex in a school where the rules of engagement are typically around management of behaviour; we had to navigate and respect the school, but also/simultaneously enable the young people to speak and have a voice.

During one poetry session with a group of young men, things were typically boisterous; the poetry classes take place during a time when the students would normally be taking another class, and in a different classroom (a Spanish-language one) than they are used to being in. In this sense, the expected order has been dismantled already and, as such, the students are out of their comfort zone. When we are somewhere unexpected, the natural reaction is to make noise and to stake our claim on that space (think how many of us shouted into the mouth of a dark cave as children in order to hear our own voices echo back and reassure us that we were indeed there, and wholly safe).

Towards the end of the lesson, one of the students was asked to leave the room by the teaching assistant who sits in the room to provide pastoral support; he was given a talking to outside and then, when he came back in, was made to stand by my chair and apologize to me before being allowed to sit back down. Such occurrences and methods for controlling behaviour are common in classrooms.

Reflection

What struck me as odd was that the general behaviour of the boys was loud but not disruptive; everyone was still writing and thinking, and they were often just vocalizing rather than internalizing their ideas. What's more, the boy who was forced to apologize to me was not behaving any better or worse than the collective he was a part of – but he was the one who was singled out – confirming a role that he no doubt already had within the school: that of the disruptive boy who leads others astray and often goes too far. So despite being in the new space, in a new class, with someone new, the natural reaction is to make noise and to stake a claim for himself; he carried the baggage of how the school already perceived him.

This was experienced across the workshops: the school had chosen the pupils for us. They had divided them along the lines of academic ability, into a group of conscientious girls who lacked confidence, and a group of mainly male students who were considered to be less academically able. In order to survive on a day-to-day basis, schools will create defined roles for each pupil and member of staff in order to weave an everyday fabric that will ensure minimal surprise throughout a week. So, pupil A is always quiet and works hard; if they become disruptive then the everyday falls apart and, thus, something must be triggering this behaviour and attention can be given to it. Or, pupil B is always disruptive and loud, they will consistently cause problems and, thus, will have energy and focus directed towards them without teachers having to reconceive the everyday behaviour of the pupil each time they disrupt: the teacher can fall back on the preconceived, everyday narrative in order to deal with their behaviour.

The school has, therefore, a set of characters. One of the key features that the space of co-production can open up is the ability to rewrite these defined roles. Tyrone (pseudonym), another pupil from the group of perceived 'difficult' boys, was captain of the football team and, thus, must have been seen to have some leadership potential. However, within the school, he was seen as angry, disruptive and unpredictable. This behaviour was connected to a recent family trauma that Tyrone was living through at the time. He left the session in question empowered and excited about the creative potential of poetry.

Similarly, the next session of the day was due to be music, and Tyrone asked if he could join in, or at least have a go at using one of the drums for the session. A teaching assistant present in the room refused, telling Tyrone he couldn't join in, or have a go at using one of the drums for the session. Tyrone was recast back into the role that the school had constructed for him in their everyday. The co-produced space allowed Tyrone to find a new space for himself, but in the wider school, this was still not a possibility for him.

Thinking about the idea of the 'art' of the classroom (Springgay and Rotas, 2015) and the acceptance of the affective (Ehret, 2018), here we recognize how our work disrupted the 'casting' of the young people but also worked within a frame of who, in one sense, 'owned' the young people. School became a process of relearning how to be and 'who to be'. In different ways, our project questioned the 'being' of the young people. As Neve, a young person on the research group, commented in her poem:

The school where I studied

I passed by the school where I studied
Most of my life and I said in my heart:
Here I am today and didn't stop.
All my life I have loved growing up and
Learning different things
I wish I could go back in time and re-do all my mistakes. (By Neve (pseudonym), from 'The School Where I Studied' by Yehuda Amichai[1])

School is a place where young people are characterized in particular ways. In our final discussion about the project with the headteacher, we found ourselves in a tussle about who 'knew' one of the participants the best – the headteacher won. Who tells stories about children and in what ways, is part of the affective 'tussle' of school. We were, however, anxious not to demonize the school. Our project was more to reframe it.

Reframing the school

Because the school had chosen the pupils for us, and had already divided them into levels of academic ability, we were aware that our preconceptions of the young people taking part in the project were already being determined

---

[1]   www.poetryfoundation.org/poetrymagazine/poems/40662/the-school-where-i-studied.

for us. We wanted to get to know them outside of these categorizations. We wanted to know who they were and what they thought of their school. Pursuing co-production within this context enabled us to reposition knowledge and power: choosing from the outset to value and situate the knowledge, thoughts and opinions that the young people held about their everyday experience in school at the heart of the research. This started by collaborating with the young people to sense-make and 'reframe the school'.

Critical incident 3: 'Photography and "reframing" the school' by Vicky

The first session in a school is always a bit awkward. The young people don't know you and you don't know them. Year eight is also the age when things start to become cool and not cool – would we be cool or not cool in their eyes and would that make a difference to how they took part? In reality, being cool doesn't matter; what does matter is finding a way to connect in order to start to build an equal relationship where the young people know that they are respected and valued, and that this project isn't a project that's just being 'done to them'. This is why co-production is important.

Co-production done well repositions who holds power; it enables young people to have agency within the project, to participate as themselves, and to place their knowledge at the centre. This is why pursuing an agenda of co-production has been so important: it generates a different 'living knowledge' of social cohesion within the school context (Facer and Enright, 2016). In working with the research team, we wanted to find out what they thought about their school. What they liked and disliked; what they thought about the physicality of the building; what school meant to them. The young people used creative tools such as photography, filmmaking and audio recording to document what they thought and felt.

In one session, the young people spent some time showing us around the school while photographing their favourite spaces and places. We asked them what they thought 'social cohesion' was within the school; they said it was making friends, having safe and favourite places where they could be together, and being able to be themselves. The young people explained why they had taken the pictures that they did, and talked through what the different spaces in the school meant to them. For one young woman, the library was an unsafe space. She told us that when she was in Year eight, she had overheard other young people saying not very nice things about her and now she doesn't like going into the library. One of the young men explained that he didn't like the library, as it meant he had to read. He also said that the library was a safe place because it was quiet and calm.

The canteen was both a safe and unsafe place. It was safe because there were counters and lots of teachers and dinner ladies about, so it was watched.

It was unsafe as lots of fights happen there and it can be very busy and loud. All of the young people liked the music room. Our young female researcher played piano, two of the young male researchers also played instruments, and one of them explained that he plays the drums as this helps him when he is frustrated and angry. Making a rhythm helps him to calm down and get into a different headspace. Another young researcher said that he likes it when everyone plays their instruments together and is part of a band making lots of noise. Each of them said that music was a way of communicating, of using a different language to work together.

Reflection

Through this activity, the young people were able to start to co-create a common language and way of working together. It also positioned their views at the centre and they set the agenda. This enabled us to collaborate with the young people in a different way, and established a resistance in us making assumptions about what they may or may not think about their lives in school. Our initial lesson plans became less urgent than a focus on co-creation.

When we think about our work, we reflect on the ways in which it did or did not contribute to an imaginary ideal of 'social cohesion'. This social imaginary had become very strong in the UK, divided by the vote to leave the European Union, and under strain from the far right and rise of the nationalist UK Independence Party there was a need to find a common language that supported civic participation and engagement. Our project was part of this move to create softer, kinder spaces for this debate to happen. The ordinary, however, overtook us.

Critical incident 4: 'Social cohesion in action' by Andrew

Walking back home from the tram stop after one of the school sessions, I pass a recent car accident on the main road. The emergency services haven't come and there is a predictable queue of traffic built up on either side of the incident, but nobody seems impatient. I notice more cars pulled up along the side of the road; people have stopped to help, and two people are standing in the middle of the road as a way of halting and directing traffic until the emergency services arrive. I'm struck by people's capacity to solve problems on their own, as a community, and I'm reminded of something that happened in the school earlier today.

The overarching theme of our work within the school has been around community cohesion; for one of the poetry exercises we were doing, I split the table of students into pairs. I put two girls with each other, and there was a moment of silence before one said, very calmly and without aggression, that they didn't get on. Very quickly the class solved the dilemma for me, with a couple of other students offering to switch partners. There was no hostility, no meanness, just a very mature and grown-up way of dealing with a lack of cohesion; in a practical way which perhaps writing could never have demonstrated.

Reflection

One of the most difficult and contested places within a school that has been deemed to be 'failing' by the UK inspection regime, OFSTED, is where power resides, who holds it, and how it might be recentred to the students and the wider community. So OFSTED, coming in with their 'knowledge', have told the school something, in a very direct power relationship. The school's reaction, in this particular case, has been to put up walls (metaphorically, if not literally), which would restrict the possibility of this kind of thing happening again – the school can retain its knowledge and its position of 'expert' if it resisted others coming in to mitigate that role.

The logic of this, certainly from the school's position, can be understood, but what it meant in terms of our work within the school, as artists, is complicated; the school held all the knowledge in the space, and were (rightly, I think), suspicious of us as university outsiders. The school, thus, had an instrumental view of the artist and their potential role within the class, our role was not to disseminate knowledge or ideas across the school as a whole (say at a training day for teachers) but just among our cohort of students on our given days with them.

What this did allow us to do, though, is enact a process of recentring which could act as a model for the wider school. I was able to decentre myself as someone with any knowledge and hand over to the students, allowing them to guide the sessions. What I suppose I'm saying about the situation with the two girls is that, in that particular moment, there was no need for any expertise; without the intervention of myself, or a teacher, the students were able to figure things out for themselves.

Our work then, was about the struggles to occupy a space, to move towards a 'desiring' pedagogy that moves beyond fixity into a movement-oriented position. This echoes Eve Tuck's (2010) critique of many educational positions as being about 'fixing' communities that are placed in deficit, as they are not doing something 'right'. Instead we moved into the 'what if?' realm of the world, where it was possible to amiably not 'get on' but make sense of that for ourselves.

Critical incident 5: 'Failing in a failing school' by all of us

When we returned to the school after Christmas in 2018, it was with a sense of dread. We were tired of not being recognized by the school entrance lobby welcome team and of coming into our room that was never fully acknowledged to be 'our room'. We were tired of writing lesson plans. Nothing seemed worth it. Andrew wanted to stop. Kate was exhausted. Vicky didn't think we had co-produced anything. We decided instead not to have lesson plans, but just see what happened. Andrew came into the poetry class and asked the young people to think about who they would like to be in five, ten, a hundred years' time. He then asked them to write instructions for how you would teach others to write poetry. Chantelle, Tyrone and Sofka led and co-developed ideas for the teaching of poetry and then proceeded to run the lesson. Here are Tyrone's instructions:

How to teach poetry
 By Tyrone

1. Ask about hobbies/interests
2. Deliver an example
3. If they are less confident, then you could give them some sentence stems
4. Use the hobby or interest as a title
5. Start with some description of your surroundings
6. Try and put some action into your piece
7. Some sort of success or achievement
8. End with some more description

Tyrone got up, wrote a poem on the board, and enacted the teaching, asking the class to contribute to a group poem:

Football

The studs messed up the grass.
My body filled with temptation as I tried to get
My ball in the back of the net.
Eyes focused, people screaming my name, as
my foot lifted off of the ground.
Outrageous sounds lift me off of the ground.
As the ball missiles into the top right corner,
Lifting the glimmering trophy, it shimmered into
my eyes as it enlightened the roaring stadium.

Reflection

Our refusal to plan the lesson and by giving the class the space, created a new kind of possibility. The teaching assistant who observed this class remarked that this kind of teaching never happened anymore. In relation to the teaching of English, Tyrone as teacher becomes the enactment of a new kind of improvisation, by which knowledge and power is repositioned within the classroom. This also meant we had to give up on our expertise as well as the role of teachers. We had to live with the uncertainty of this process. We learned to embrace the 'queer art of failure' (Halberstam, 2011). What this taught us was that the giving up of practice lets in new ways of knowing and being.

8

Unplanning

*Kate Pahl, Richard Steadman-Jones and Lalitha Vasudevan
with Steve Pool*

The projects discussed in this book represent work in which one must feel one's way. One has to find the path as one travels. Some of what Robin Nelson says in *Practice as Research in the Arts*[1] sheds an interesting light on this issue. At a number of points, he talks about the importance of a *'clew'* to 'Practice as Research' projects.[2]

According to the *Oxford English Dictionary* (OED) definition 2a, a 'clew' is a ball formed by winding thread or yarn. Definition 3a notes that in various mythological or legendary narratives (especially that of Theseus in the Cretan Labyrinth), 'a ball of thread' is mentioned as the means of 'threading' a way through a labyrinth or maze; hence, in many more or less figurative applications the term denotes that which guides through a maze, perplexity, difficulty, intricate investigation and so forth. When Theseus entered Daedalus' Labyrinth to slay the Minotaur, the king's daughter – Ariadne – gave him a ball of thread, which he unwound as he went. In this way, he avoided getting lost and was able to find his way out of the maze once he had killed the monster.

This image acknowledges that one can easily become 'lost' in a practice-oriented research project. Or perhaps it is better to say that one can lose the sense that what one is doing constitutes what Nelson calls a 'research enquiry'. If one is really using a practice as a research method, one has to

---

[1] Robin Nelson (2013) describes research in the arts where the research exists as practice. This is not quite what we are talking about here, but there are elements of this approach, including a focus on doing and embodied knowing, that we think are helpful.

[2] Nelson (2013) talks about how the old form of the word 'clew' literally denotes a thread, and students have found it to be a productive metaphor for holding on to the line of the research enquiry as it weaves through the overall process.

accept that there are things that can't be thought out in advance because it is only by *doing* that one will make discoveries. But to maintain some sense of direction – or orientation – one needs a 'clew' which can guide the process of learning-from-practice. The 'clew' is whatever one uses to orient the practice in the direction of learning.

The image of the ball of yarn is also useful because it suggests that the 'clew' is something improvised. This is not about a highly formalized research methodology – it is about something that allows one to make good decisions in the moment and respond constructively to what unfolds in the here and now.

There is a connection between the use of the 'clew' and the process of documentation. Theseus marked his path by laying down a trail of yarn, unwinding it from the 'clew' as he went along. To document practice is, in a sense, to mark out the path that one has travelled so that one can at some future point retrace it. If practice is a labyrinth, then guiding the process of exploration is an improvised device which one lays down as one goes and returns to as one emerges from the work. Doing so is not driven by a desire for replication or generalizability, but rather to provide a tether in the midst of feeling unmoored, such that 'lost' is not a debilitating condition.

What do you do if you are lost?

Feeling lost is rarely comfortable; yet, if we are to continue to explore at the edges of what is known, it is not only inevitable but also essential. Getting lost in research or in the world is rarely spoken of as a good thing. However, it is not something that should be avoided at all costs. Being lost can be described using more positive terms such as 'indeterminacy', or 'unknowing'; yet, none of this happens without first stepping into the woods, disregarding the well-trodden paths and forging recklessly into the dense thicket. The clew helps with getting lost in that it provides some orientation despite the fact you are treading in an unknown place. We can think about the provision of clews in terms of the concept of unplanning, which is neither 'not planning' nor rigidly prescriptively planning but rather something in between – a kind of open form of planning which allows space for the unexpected and the serendipitous.

There are two aspects to unplanning. One is the giving-up – not being wedded to research methods, an orientation of openness to the unexpected. In this sense, unplanning can be seen as an unravelling of our expectations for the future. To enter on a journey with no intentional destination can feel vague, or imprecise, or perhaps not a journey at all. It is certainly not what normally gets described as good research practice. Unplanning emerges as a process, a curatorial choice that opens into a dappled clearing in the dark woods where something unanticipated has the potential to emerge. The idea of 'giving up' control of our expectations doesn't necessarily result in chaos.[3] Secondly, a process-oriented approach to unplanning is an open form. To unplan is to fine-tune approaches that take account of interruptions that have not yet happened. This involves interrupting the tracks of thinking or doing that are established and ill-thought habits, and to decentre research custom and practice to create new configurations.[4]

The root of the word 'curate' is 'to care' (Latin *curare*). Thinking of this work as 'curation' rather than 'research' then enables us to refocus on the importance of caring. This is not only to care for the people and ideas involved but also, more importantly, to care about the journey and the process. To care for a future that is open is to accept that knowing anything can only ever be temporary. 'Living knowledge' (Facer and Enright, 2016) is not something that can be captured or even contained.[5] Yet, the pursuit of research remains vital.

There are times when a good plan and an excellent way to communicate it to others are the only possible options. For example, in order to plan a national response to a pandemic or even to organize a party for a seven-year-old, to set out without a plan would be reckless. However, the success of planning in achieving known outcomes can sometimes trick us into thinking it will be equally applicable to situations where the desired outcome is completely obscured from anyone's view. What is thought to be a good outcome might not look like good practice; it might look like chaos in classroom terms. For example, when working with young people, planning needs to balance their need to play, dance and mess about.[6] By sitting with this work as work that is undefined, different outcomes can open out.

---

[3] For example, Stephen Willats's (2012) *Artwork as Social Model: A Manual of Questions and Propositions* provides a number of ways in to an unplanned system of making art, including, for example, informal networks which are not designed to be a part of any plan, but can lead to transformative action.

[4] *Transmissions* edited by Kat Jungnickel (2020) provides a number of examples of decentring of research traditions, including a Chapter 5 ('Exchanging'), where the reader becomes a co-author of the text.

[5] See the 'Creating Living Knowledge' report on the Arts and Humanities Research Council (AHRC) Connected Communities programme. Keri Facer and Bryony Enright (2016) describe this approach in the Connected Communities programme and highlight their distinctiveness. What is noticeable about these collaborative projects, however, is that the theories and concepts that are being developed are not detached from the conditions of their production They write, '[t]he theory that is being built is a form of "living knowledge" a praxis knowledge that connects lived experience on the ground with the wider body of national and international critical knowledge'(p 134).

[6] In our article, ' "Living your life because it's the only life you've got"', 'messing about' was something the children identified as an important part of school experience. However, it was hard to identify this within formal education contexts, except for the dance class run by the children (Pahl and Pool, 2011).

Within the work, it is the 'moments that emerge' that for some reason gain momentum and hold the real potential. However well-planned one's research is, the interesting thoughts that flash into focus never happen on the well-trodden paths; rather, they present themselves to the researcher in the thick brambly undergrowth as we try and cut a pathway. The clew is a way to orientate yourself, but it is a provisional way to orient yourself. This path sometimes leads to a clearing, but more often, it takes us deeper into the forest. The trick is to learn to live with the sensation of having no fixed abode, to get more comfortable with feeling uncomfortable. The purpose of an unplanning approach is to challenge persistent hierarchies and to challenge conceptions of 'best practice'. It affords an approach that values a meshwork of ways of knowing and doing. This opens potentials that are not yet known or even desired. It supports new configurations and stories. The unplanning that we describe here is familiar to many people in the arts.

There are different cultures and modes of being and doing in which unplanning is found. Here are some examples that musicians and artists have employed to engender spontaneous work.

Fluxus was 'an international network of poets, artists and composers who worked across different media, and who sought to integrate art into everyday life.[7] Fluxus can be understood in the context of a movement with 'an alternative attitude towards artmaking, culture and life'.[8] The network was working primarily in the 1960s and 1970s. Artists associated with Fluxus tended towards a radical stance and were loosely connected; they included John Cage and Yoko Ono. James Miles and Stephanie Springgay (2020) write:

> Importantly, Fluxus artists made artwork and art events as curricular materials and as pedagogical scores for their classrooms. There is a distinct blurring between curriculum, pedagogy, and art in the Fluxus

---

[7] Miles and Springgay (2020) give a useful background to the Fluxus movement, drawing on an archival project undertaken at the Ontario Institute for Studies in Education (OISE) as part of a wider project called the 'Pedagogical Impulse': https://thepedagogicalimpulse.com.

[8] An example of this work included the 'Instant Class Kit' described as follows: 'Instant Class Kit' is a portable curriculum guide and pop-up exhibition dedicated to socially engaged art as pedagogy. Produced as an edition of four, the kit brings together contemporary curriculum materials in the form of artist multiples such as zines, scores, games, newspapers and other sensory objects from a diverse group of artist-educators across North America. 'Instant Class Kit' is closely modelled on the multi-sensory and open-ended strategies of Fluxkits, as well as hands-on learning kits commonly used in K-12 education. Combining these influences, 'Instant Class Kit' offers an interactive and speculative approach to teaching that is participatory, collaborative, and social justice oriented. https://thepedagogicalimpulse.com/project-description/.

movement, and it is this 'intermedia' practice … that is central to their influence on the current art education field. (p 1009)

Fluxus artists found ways of creating disruptions or provocations within pedagogical encounters. For example:

This led to Flux editions such as George Brecht's Water Yam (1963), a cardboard box housing an assortment of printed cards in various sizes that contain abbreviated, haiku-like prompts called 'event scores.' These open-ended instructions invited participants to enact everyday actions or contemplate impossible scenarios. This eventually morphed into what became known as Fluxkits, which often included a wide range of objects and provocations created by Fluxus artists often contained within a suitcase or box. Fluxkits were produced in multiple editions, as part of anthologies, and for distribution via mail order. (Miles and Springgay, 2020, p 1010)

These methods were taken up in education in a variety of ways, providing a means of engineering spaces where surprise was part of the encounter with the world.

One of the artists who has been prominent in these projects, Steve Pool, trained us as academics in a mode of working which he talks about in terms of 'dogme'. When we first started working on 'Writing in the Home and in the Street', he used to talk about the idea of setting oneself some rules and then following those rules rigorously wherever they led. So, for example, we (Richard and Steve) set ourselves the goal of walking round Rotherham, UK one brutally cold January morning and taking exactly one hundred photographs of written texts (described in Chapter 2).

In talking about these sorts of projects, Steve often used to allude to 'Dogme 95' –the approach to film-making that Lars von Trier and Thomas Vinterberg formalized in the Dogme Manifesto (also known as the 'Vows of Chastity').[9] This is a set of constraints that film-makers who subscribe to the manifesto must respect, a framework of rules that serves to amplify creativity rather than curtail it. The rules of Dogme 95 are as follows:

1. Shooting must be done on location. Props and sets must not be brought in (if a particular prop is necessary for the story, a location must be chosen where this prop is to be found).
2. The sound must never be produced apart from the images or vice versa. (Music must not be used unless it occurs where the scene is being shot.)

---

[9] This is where the original Dogme is posted: https://pov.imv.au.dk/Issue_10/section_1/artc1A.html.

3. The camera must be hand-held. Any movement or immobility attainable in the hand is permitted.
4. The film must be in colour. Special lighting is not acceptable. (If there is too little light for exposure, the scene must be cut or a single lamp be attached to the camera.)
5. Optical work and filters are forbidden.
6. The film must not contain superficial action. (Murders, weapons, and so on must not occur.)
7. Temporal and geographical alienation are forbidden. (That is to say that the film takes place here and now.)
8. Genre movies are not acceptable.
9. The film format must be Academy 35 mm.
10. The director must not be credited.

As well as constraints, there can be propositions or provocations that give you guidelines about what to do and not just what to avoid. This leads on to the idea of open scoring. In the 1950s, composers such as John Cage experimented with open scores that provided information to performers in a new way, not necessarily with traditional notation but other forms of media. These could be interpreted in a much more open way to classical music scores. This offered a form that was indeterminate and offered a variety of paths through a piece.

Indeed, the chapters in this book explore the 'variety of paths' or various stages of thinking along the way of doing research, as well as its embodied and enchanted elements. In that sense, the experiences described in this book are the 'headnotes'[10] that we carry with us of research encounters. Simon Ottenberg (1990) talked of the interrelation of fieldnotes and 'headnotes' and how these headnotes change as we grow and shift. When we do a research project, it evolves. The thinking starts in one place, and experiences accrue within the project. The work is, therefore, layered and memories also layer the work. This book is an entire headnote that speaks to other people about the process.[11]

In the 'Worldizing' chapter (see Chapter 3), we explored what happens when you are guided by the sites and spaces of the projects. We argued that tangible and intangible forms of knowing got tangled up – some of these having antecedents in theory; others not. This fuzziness could be

---

[10] Headnotes are internal memos that accompany the making of fieldnotes by an anthropologist (Ottenberg, 1990).

[11] This conversation about headnotes and footnotes was supported by the late Brian Street, who taught us to see that the eternal reflections from fieldnotes were as valuable as the fieldnotes themselves. See also Sanjek (1990).

understood as central to projects that were conducted in everyday settings, as they wended their way across the everyday, and the work of the projects became something held by many people. Within this mix, though, we sometimes needed the 'clew' of theory to guide us through something. For example, habitus became a 'clew' for Kate as she became caught in the intergenerational narratives of homes and their meanings (see Chapter 3). The idea of 'worldizing' was to provide an expression for this sort of thinking and doing. The theory could come in and out of the picture, much as car radios do in streets where it is hot, and people are playing music. The trick is to listen when it is useful, but not all the time.

Chapter 4 is concerned with worthiness. To recognize the worthiness of another's experience is, in a sense, to turn away from research methods that instrumentalize that person. In this way of working, the people who join the project are not 'subjects' or 'informants', their contribution to the research carefully encoded in the research design. They are ends in themselves and they are fully involved in determining what the research will become. As such, the ethical imperative – to see the worthiness of others – requires an approach characterized by unplanning. In a sense, this is another renegotiation of the status of theory. Theory does not drive the project. People do.

Enchantment can be seen as resistance to rationalization and bureaucratization. As such, it is a valuable outcome to a programme of work. However, one cannot specify re-enchantment as an output of a project and move towards it step-by-step. In Chapter 5, we examined how re-enchantment took place as a process of writing and thinking. To look at the parallels between language and magic was to reenchant language.

Chapter 6 is about embodied ways of knowing: an image that we return to often is that of 'feeling your way'. And feeling your way is an embodied way of proceeding. It involves tactility. You reach out your hands to touch what is in front of you in the darkness. The body inches forward. You put your feet down carefully in case there is a step or an obstacle or a patch of soft ground. But this isn't just a metaphor. The projects described here involved inhabiting spaces with others in an embodied way and we had to feel our way through those literal spaces.

In Chapter 7, we examined the use of hypertext as a means of gathering and presenting the multiple voices that spoke within a project. If you allow voices to speak without mediating them in any strongly interventionist way, you de-emphasize your own plans for the project and work with what your collaborators volunteer. What's more, the hypertext offers multiple pathways to readers, each of them bringing together a unique configuration of voices. Each reader's experience is different and the author of the work relinquishes control over what any individual reader will encounter. Hypertext creates a new ontological setting for statements. It is almost labyrinthine. And this takes us back to the idea of the clew – the means by which Theseus finds his way through the labyrinth.

Being vulnerable in the projects

Undergoing research with young people, alongside them, is unnerving work. Rendering oneself vulnerable to young people, to adults in their lives, and to the communities where they live is itself an act of rebellion against the normative expectations of research. But it is work that is necessary in order to undo the embedded legacies of methodological malpractice that have viewed and treated communities as disposable commodities. So, we move forth, slowly and cautiously, with participatory impulses in our hearts and bags of materials to convey this spirit in practice. We enter classrooms, multi-purpose rooms, churches; we approach riverbanks, coffee shops, playgrounds, with a familiar sense of the possible clashing with a nagging worry (about what may seem, at first, impossible or insurmountable). How will our invitations be received? Come collage with us! Let's redraw the knowledge lineages to be inclusive of everyone's stories! Who wants to make a movie?

We seek signs of recognition that our new collaborative partners view their roles as ones that are meant to help us create the spaces for enquiry. But communities and people who are often at the centre of research – whose experiences we have sought to centre in our work – have also had a long history of having their trust in official institutions and groups fractured and broken (Patel, 2014, 2016). Suspicion towards research that seeks to be inclusive, participatory and often lacking a concrete plan makes sense; all forms of knowledge are not the right of everyone (Tuck and Yang, 2014). Why should people trust us? Can you begin at trust? Can you offer trust as an opening gambit? And when trust collapses, and your impulse is to run and keep running, what will allow you to stop and return to offer trust and build trust again? Collaborative research does not work without it.

A digression

Train journeys can be great places to think and plan. They offer a number of opportunities for unplanning as well. These include lost tickets (often a problem), lost platforms and, in some cases, lost trains. Trains can also go on digressions and diversions. Trains are a space for conversation among strangers, colleagues, friends over a gin or beer or tea from the buffet. While train journeys are not like research projects, they have some similar qualities; they are purposeful, have an intent, but offer unexpected experiences along the way. Maybe our research journeys should also recognize the little things in the interstitial spaces – the shared hunt for misplaced tickets, the saving of a seat, the recognition of a need to rest, or a laugh at the end of a long day.

Train journeys are a form of transportation, yes, but on more than one occasion, they have served as chariots and respite away from the site of fractured trust. Lalitha once fled from a research site towards the high-speed train that would carry her from New York City to Philadelphia, to the home where she had lived for over a decade. On that train ride, proffering comfort as its wheels moved swiftly along painstakingly laid tracks, I (Lalitha) contemplated the scene that had played out in the moments before my decision to leave the city a day earlier than planned. I had been engaged in participatory ethnographic work at Journeys (see Appendix) for a few years at the time and had developed good working relationships with the young people and staff there. So when one of the teachers, in a scheduled meeting to discuss the next cycle of the digital media workshop, called into question its very purpose and the pedagogical performance of me and my research assistant (who had taught the workshop with me), I was caught flat-footed. The critique was not the issue; it could easily be addressed with some discussion about participatory practices in pedagogy and research – and learning to be more explicit about these practices, lest miscommunication allow ill feelings to fester, was a good thing. No, the worrisome bit was the implication that in designing and facilitating such an experience, and ultimately allowing the young people to lay bare their creative productions in a screening with an audience of Journeys staff and their friends and family, I had somehow put the young people at risk. At issue was the *lack* of censorship we had practised in our workshop, which itself wasn't *verboten* but *became* an issue when the final products – the short films and videos the young men had created over the course of ten weeks, some of which contained 'objectionable' material – were shared in a semi-public venue.

In that moment I experienced a form of dread, specifically researcher remorse. The 90-minute train journey between New York City's Penn Station and Philadelphia's 30th Street Station remains vivid in my memory and the biggest learning was a simple one: we had not done enough to prepare our audience to receive the films. We *had*, in anticipation of possible misinterpretation, produced a 'behind the scenes' video in which the young people's processes and deliberations were explicated in a documentary-style format. However, in a desire to place the young people at the forefront, we had shared our film last and, in doing so, we squandered the opportunity to prepare the audience.

In Chapter 2 we discussed the importance of preparing an audience. The truth is, as researchers who have entered into a covenant of collaboration with others, we are also members of the audience of the unfolding activities and experiences that comprise research co-production. Preparation depends upon unplanning, and recognition of that failure is inevitable and perhaps even necessary. This is wrapped up into the poetics of research.

Failure

In the midst of this work, it is of course possible to fail. Failure has positives in that it can be a critique of the current conditions of how we live. Jack Halberstam (2011) has talked about the idea that the negative affect associated with failure might form part of a critique of contemporary capitalism. Not knowing and unknowing are likewise spaces that open up different possibilities (Vasudevan, 2011a). In her book *The Social Life of Nothing*, Susie Scott (2019) describes 'nothing' as 'immense, chasmically wide and deeply meaningful' (p 1). No-things can 'jolt us into life' and enable trajectories that can open up a landscape of lost-ness and disorientation, feeling our way into a new landscape. Vulnerability creates its own opportunities (Weber, 2017). Letting go is part of the process of opening up to new and strange worlds, but something then is also lost. Carl Jung's concept of the trickster appreciates that 'all the foolish, clownish mistakes we make are necessary foundational experiences on which to build a fuller, integrated sense of self' (Scott, 2019, p 161). But the moment of failure can also open up new vistas, as Sarah Truman et al (2021) articulated in a celebration of the act of saying 'no':

> We would like to enter the space of no, the space of the student who falls silent, who says no to participating, as well as the teacher who leaves. These students have too long been consigned to the bin of academic writing to privilege the good girls who … produce poetry on time. We are interested in the field of no as a porthole to an alternative universe, a TARDIS (a nod to Doctor Who) that rocks with the intelligence of another kind of resistance. This is the beginning of many more noes. (p 12)

Halberstam's (2011) book, *The Queer Art of Failure*, aims 'to think about ways of being and knowing that stand outside conventional understandings of success' (p 2). In the work we do (for example, 'Reimagining Futures', 'Imagine'), the projects sought to do work with children that lay outside schooling in in-between spaces. While some of these projects didn't work in a 'schooled' sense, they produced affective moments that continued to linger in our minds. Halberstam's book 'dismantles the logics of success and failure' (p 2), which are current in our society, suggesting that 'failing, losing, forgetting, unmaking, undoing, unbecoming, not knowing' may 'offer more creative, more cooperative, more surprising ways of being in the world' (pp 2–3). In this sense, '[f]ailure preserves some of the wondrous anarchy of childhood and disturbs the supposedly clean boundaries between adults and children, winners and losers' (p 3). And while, 'failure certainly comes accompanied by a host of negative affects, such as disappointment, disillusionment, and despair, it also provides the opportunity to use these negative affects to poke holes in the toxic positivity of contemporary life' (p 3) (see Introduction in Halberstam, 2011).

Relevant here is a short essay that Halberstam published in 2012 with the title 'Unlearning'. It appeared in the Modern Languages Association (MLA)'s journal, *Profession*, and it is very much a 'state of the humanities' paper. It constitutes a response to a 'defence of learning' articulated in the same year by the then-president of the MLA, Russell Berman. Halberstam contends that, while learning may be important, 'learning to unlearn' is also crucial:

> [L]earning ... how to break with some disciplinary legacies, learning to reform and reshape others, and unlearning the many constraints that sometimes get in the way of our best efforts to reinvent our fields, our purpose, and our mission. Unlearning is an inevitable part of new knowledge paradigms if only because you cannot solve a problem using the same methods that created it in the first place. (Halberstam, 2012, p 10)

In co-production, one of the challenges is trans-disciplinary work, where to work together, you have to 'unlearn' some of what is in your discipline to listen to the other person. In this context, Halberstam returns to an idea that he elaborated in *The Queer Art of Failure*: the need to pay attention to '"silly" archives' (2011, p 19) – texts that 'do not make us better people or liberate us from the culture industry' but which 'might offer strange and anticapitalist logics of being and acting and knowing' and which 'harbor covert and overt queer worlds' (pp 20–21). The idea of 'silliness' has been developed by Peter Kraftl in his book *After Childhood* (2020) as a way to describe how things intersect and are juxtaposed in complex and non-rational ways, disrupting conventional ideas of the developing child.

In 'Unlearning', the text in question is *The Muppets*, a movie which was released in 2011 and of which Halberstam says:

> In the film, the Muppets teach us many lessons and stress the importance of unlearning many more: some have to do with the importance of holding on to older technologies of representation, like puppetry, in the face of blazing new feats of engineering, like computer-generated imagery (CGI) animation; others concern the magic of certain often-debased forms of theatricality in an age of competing frames of reality. Still more lessons from the Muppets lead viewers to unravel the origins of queer self hood from within the confines of heterosexual childhoods.

> The Muppets focus our attention on the power of collectivity in the face of corporate greed, the importance of nonhuman love, and the subtle differences between men and Muppets. (Halberstam, 2012, p 14)

Three ideas from this chapter come to the fore. The first is the idea of failure as a productive space of practice. The second is the importance of unknowing as a stance within interdisciplinary projects. The third is the need to systematically unplan as well as plan. We have learned to accept that failure is not the end of the world but something to be shared within a thesis or a book, or with a colleague or a friend. Through unknowing, we have also learned to appreciate complexity and we have learned that polyphonic discursive practices and unplanning can enable a more complex intellectual praxis to evolve. Taken together this enables a more diffuse concept of knowing to take root.

Conclusion

The stance within this work is one of risk with acceptance. The Muppet Gonzo seems to personify this stance with his daredevil stunts, all of which end disastrously. Gonzo plunges into dangerous situations and illustrates the value of the non-instrumental chaotic person. In doing so, he also reveals the wisdom within such a posture. Our aim in writing this book was to investigate the qualities of an approach that Gonzo might take to research. This does not mean we act in a fool-hardy or risk-taking manner, but it does mean respecting the complexity of projects which might appear chaotic. We recognize that the dare-devil character is part of the mix of learning and making – it is one of the qualities of a research encounter. For example, in collaborative interdisciplinary projects, artists as practitioners can engage with such characteristics to produce insights for the project team, for example in a drama workshop where uncertainty is built into the practice. Within research encounters, qualities such as unlearning, failure, chaos and risk-taking are embedded in a practice that attempts to go close to where people are, and this shouldn't be hidden when we write up the experience.

As we noted earlier, we wanted to elucidate the 'headnotes' – those that are present in the moment, and those that emerge upon reflection – when engaging in a variety of collaborative research relationships with young people, teachers, artists, scientists, family members and others. But far from being an exercise in navel-gazing, this internal excavation was done with an eye towards mapping new and sometimes unfamiliar terrain in the landscape of 'research *with*' – terrain that is not covered in methods books. To do so, we dwelled in the messiness of research and we sought to identify guideposts for fellow wanderers as they – you – embark on future endeavours that are challenging, a little scary, and that plunge all parties involved into the depths of enquiry without any certainties about where it may lead.

This chapter has described the journey of research practices as a process that can be complex, and can include digressions and diversions and non-linear narratives. This approach to collaborative research is not without structure or discipline, and we sought to attend throughout our recollections and re-presentations to the variation in how structures and disciplines are manifested in, around, and in between collaboration. But, the dare-devil who isn't frightened of failure has possibly more wisdom to offer than we might imagine.

Researchers might like to find these ideas useful when planning projects:

- The concept of being lost can be productive within research spaces. Feeling one's way is embedded within collaborative research but it is an intentional mode of working together. Our idea of the 'clew' provides a sense of a structure needed to support the process of being immersed within research. The ball of thread provides meaning to the unfolding of research.
- Giving up research plans can be helpful in that this leads to further discoveries and can also create new kinds of structures in which research can happen. The indeterminacy in the work is as important as the final product.
- The experience of doing this kind of research can feel uncomfortable and can decentre research expertise but it is an important aspect of the project.
- Working with art and artists and necessarily blurring disciplinary boundaries can enact some of this work. The concept of dogme and the structures of art practice can enable innovative research practices. Constraints and affordances such as open scoring can open new avenues for exploration.
- Thinking with the projects involves employing a poetics of research. We need to think about artful approaches as a form of knowing. This can include vulnerability and listening as a mode of thinking.
- Failure is part of the story of research. When writing a thesis, or a paper, sharing stories of failure can help a reader understand what happens on a research journey. Eliding failure takes away a richness of experience that is important for others to learn about.

The chapters in this book have been about what it means to *feel* research, to experience journeys that are open-ended, like hypertext. We have sought to work *within* the spaces of the projects, rather than think about the end. This provides a different language of description for the process of doing research, which we have called the *poetics of letting go*. This is not a simple process but one that requires a sustained and coherent focus on the clew, the 'through-line' of the project, while enabling the letting go along the way. Our hope is that by valuing this feeling more strongly, the thoughts and ideas that accompany this journey can be surfaced and valued.

Notes on the Work

Cristina Salazar Gallardo

1. The bus ride after the meetings was the time for scribbling notes into my notebook, in the frenetic pace of someone who wonders if the details, the smiles, the hugs are going to be lost by the time I try to write about this.
2. I would try to come up with new words, ways of expressing the warmth of the experiences. To write 'We ate pizza together and laugh' but to sharpen the description to explain what the laughter felt like.
3. 'Roles intertwine – teacher, mother, researcher, aunt, friend, lover, sister – producing witnessings that shake boundaries' (Pillow, 2019 p 118).
4. I appreciated any time food was present in the workshops. There were multiple ways in which it changed the dynamic, and food seemed to relax everyone in the classroom. We didn't have to think about our roles. We were just sharing a meal.
5. One time, I couldn't attend a workshop and Z, one of the students, sent me a Post-it with a fellow researcher. 'Hey Cristina, we missed you today.'
6. I miss them today.
7. The day after the election, I attended a meeting with immigrant youth activists. We cried and hugged each other, in a classroom. We were so afraid.
8. I have never felt more inadequate at writing than whenever I try to write about these experiences.
9. I hope for better language, verbs and adjectives that explain the embodied experience of being within these spaces. And the pauses that live between sentences and utterances, that weave a relationship and shape our understanding of these worlds.

10. Years pass, you conduct research in other places, in other contexts, and yet, you still go back to those times whenever you are meeting participants. The way in which you sit, in which you make eye contact, and in which you ground yourself in the respect for the people in front of you are just ways of honouring the young people who taught you how to witness and be present with them.

List of Projects and People with Dates and Funders

Communicating Wisdom: Fishing and Youth Work. Research team: Johan Siebers (PI), Kate Pahl, Richard Steadman-Jones, Hugh Escott (RA), Steve Pool, Andrew McMillan, Marcus Hurcombe (youth worker). The Arts and Humanities Research Council (AHRC) funded one year (2013–2014).

Co-producing Legacy. Research team: Kate Pahl (PI), with Amanda Ravetz, Helen Graham and Steve Pool (Co-Is). AHRC funded one year (2014).

Education In-Between. Research team: Lalitha Vasudevan (PI), Mark Dzula (RA), Mathangi Subramanian (RA). Funded by New York City Office of Children and Family Services and Teachers College, Columbia University (2004–2009).

Imagine. Research team: Kate Pahl (PI from 2014), with Sarah Banks, Angie Hart and Paul Ward (Lead Co-Is). Included the 'Park Hill' project led by Prue Chiles, with Paul Allender, David Bell and Louise Ritchie. The Economic and Social Research Council funded for five years (2012–2017).

Language as Talisman. Research team: Kate Pahl (PI), Jane Hodson, Richard Steadman-Jones, David Hyatt, Steve Pool (artist), Hugh Escott (RA), Andrew McMillan, Cassie Limb (artist), Marcus Hurcombe (youth worker), Deborah Bullivant (community researcher). AHRC funded one year (2012).

Odd: Feeling Different in the World of Education. Research team: Rachel Holmes (PI), Kate Pahl, Amanda Ravetz, Becky Shaw, Steve Pool, Jo Ray (RA), with Alma Park Primary School, Manchester, UK. AHRC funded three years (2018–2021).

'*Questioning the Form*'. Research team: Kate Pahl (PI), led by Gloria Kiconco, Lisa Damon, Charity Atukunda, Su Corcoran. AHRC/GCRF funded one year (2020–2021).

'*A Reason to Write*'. Research team: Steve Pool and Kate Pahl with Gooseacre Primary School. Creative Partnerships funded one year (2010).

Reimagining Futures. Research team: Lalitha Vasudevan (PI), Lydia Browne (RA), Jeremi Calderon (intern), Devon Colwell (intern), Eric Fernandez (youth worker), Melanie Hibbert (RA), Kristine Rodriguez Kerr (Project lead and RA), Eva Neves (RA), Katie Newhouse (RA), Ahram Park (RA), Anna Pizarro (RA), Cristina Salazar Gallardo (RA), Pedro Sacaza (intern). Funded by The Robert Bowne Foundation; New York City Office of Children and Family Services; and Teachers College, Columbia University (2009–2019).

'Researching Community Heritage' also known as 'Portals to the Past'. Research team: Robert Johnson (PI), Brendan Stone and Kate Pahl (Co-Is), with Steve Pool, Hugh Escott, Marcus Hurcombe (youth worker). AHRC funded one year (2014).

'Taking Yourself Seriously': Artistic Approaches to Social Cohesion. Research team: Kate Pahl (PI), Steve Pool, Zanib Rasool, and Andrew McMillan (Co-Is), with Katy Goldstraw (RA) in collaboration with Association for Research in the Voluntary and Community Sector with Vicky Ward. AHRC funded one year (2017–2018).

Voices of the Future. Research team: Kate Pahl (PI) with lead team Caitlin Nunn, Peter Kraftl, and Simon Carr and Samyia Ambreen (RA) collaborating with children and young people to reimagine treescapes. The Natural Environment Research Council funded three years (2021–2024).

References

Abu El-Haj, T.R. (2009) 'Imagining postnationalism: arts, citizenship education, and Arab American youth', *Anthropology & Education Quarterly*, 40(1), pp 1–19.

Agrippa, Heinrich Cornelius (1651) *Three Books of Occult Philosophy*, translated by J.F. London: Gregory Moule.

Ammons, A.R. (1986) "Poetics," in *The Selected Poems: Expanded Edition*. New York: Norton, pp 44–45.

Amsler, S. (2019) 'Gesturing towards radical futurity in education for alternative futures', *Sustainability Science*, 14(4), pp 925–930.

Angelou, M. (1991) 'Ailey, Baldwin, Floyd, Killens, and Mayfield', in *I Shall Not Be Moved*. New York: Random House, pp 47–48.

Bachelard, G. (1958/1994) *The Poetics of Space*, translated by M. Jolas. Boston, MA: Beacon Press.

Baker-Bell, A. (2020) *Linguistic Justice: Black Language, Literacy, Identity and Pedagogy*. London: Routledge.

Bakhtin, M.M. (1981) 'Epic and novel', in M. Holquist (ed) *The Dialogic Imagination*. Austin, TX: University of Texas Press, pp 3–40

Banks, S., Armstrong, A., Booth, M., Brown, G., Carter, K., Clarkson, M. et al (2014) 'Using co-inquiry to study co-inquiry: community-university perspectives on research', *Journal of Community Engagement and Scholarship*, 7(1), pp 37–47.

Barad, K. (2006) 'Agential realism: how material–discursive practices matter', in *Meeting the Universe Halfway: Quantum Physics and the Entanglement of Matter and Meaning*. Durham, NC: Duke University Press, pp 132–185.

Barnes, D. (2020) *I Am Every Good Thing*. Illustrated edition. New York: Nancy Paulsen Books.

Barrett, E. and Bolt, B. (eds) (2010) *Practice as Research: Approaches to Creative Arts Enquiry*. London: I.B. Tauris.

Barthes, R. (1978) *A Lover's Discourse*, translated by R. Howard. New York: Hill & Wang.

Barton, D. and Papen, U. (eds) (2010) *The Anthropology of Writing: Understanding Textually Mediated Worlds*. Illustrated edition. London: Continuum.

Bell, D.M. and Pahl, K. (2018) 'Co-production: towards a utopian approach', *International Journal of Social Research Methodology*, 21(1), pp 105–117.

Bennett, J. (2001) *The Enchantment of Modern Life: Attachments, Crossings, and Ethics*. Princeton, NJ: Princeton University Press.

Berlant, L. (2016) *Interview with Lauren Berlant*. Singidunum University (2016 Summer School for Sexualities, Culture and Politics). Available at: www.youtube.com/watch?v=Ih4rkMSjmjs

Berlant, L and Stewart, K. (2019) *The Hundreds*. Durham, NC: Duke University Press.

Bhopal, K. (2018) *White Privilege: The Myth of a Post-Racial Society*. Bristol: Policy Press.

Bloch, E. (1969) 'Über den Begriff Weisheit', in *Philosophische Aufsätze zur objektiven Phantasie*. Frankfurt am Main: Suhrkamp Verlag, pp 355–385.

Bloch, E. (1986) *The Principle of Hope, Vols 1–111*, translated by N. Plaice, S. Plaice, and P. Knight. Oxford: Blackwell Publishing.

Blommaert, J. and Jie, D. (2020) *Ethnographic Fieldwork: A Beginner's Guide*. 2nd edition. Bristol: Multilingual Matters.

Bourdieu, P. (1977) *Outline of a Theory of Practice*. Cambridge: Cambridge University Press.

Bourdieu, P. (2010) *Distinction: A Social Critique of the Judgement of Taste*. Abingdon: Routledge.

Bourdieu, P. and Wacquant, L.J.D. (1992) *An Invitation to Reflexive Sociology*. 1st edition. Chicago, IL: University of Chicago Press.

Brockmeier, J. (2002) 'Remembering and forgetting: narrative as cultural memory', *Culture & Psychology*, 8(1), pp 15–43.

Brown, M., Pahl, K., Rasool, Z. and Ward, P. (2020) 'Co-producing research with communities: emotions in community research', *Global Discourse*, 10(1), pp 93–114.

Campbell, E. (2018) 'Methodology: an introduction', in E. Campbell, E. Pente, K. Pahl and Z. Rasool (eds) *Re-imagining Contested Communities: Connecting Rotherham through Research*. Bristol: Policy Press, pp 87–90.

Campbell, E. and Lassiter, L.E. (2010) 'From collaborative ethnography to collaborative pedagogy: reflections on the other side of Middletown Project and community-university research partnerships', *Anthropology & Education Quarterly*, 41(4), pp 370–385.

Campbell, E. and Lassiter, L.E. (2014) *Doing Ethnography Today: Theories, Methods, Exercises*. Chichester: Wiley-Blackwell.

Campbell, E., Pahl, K., Pente, E. and Rasool, Z. (2018) *Re-Imagining Contested Communities: Connecting Rotherham through Research*. Bristol: Policy Press.

Carson, A. (1999) 'The idea of a university (after John Henry Newman)', *The Threepenny Review*, 78, pp 6–8.

Chiles, P., Ritchie, L. and Pahl, K. (2018) 'Co-designing for a better future: re-imagining the modernist dream at Park Hill, Sheffield', in S. Banks, A. Hart, K. Pahl and P. Ward (eds) *Co-Producing Research: A Community Development Approach*. Bristol: Policy Press, pp 115–134.

Coffey, A. (1999) *The Ethnographic Self: Fieldwork and the Representation of Identity*. London: SAGE.

Cohen, J. (2020). 'A teenager didn't do her online schoolwork. So a judge sent her to juvenile detention', *ProPublica*, 14 July. Available at: www.propublica.org/article/a-teenager-didnt-do-her-online-schoolwork-so-a-judge-sent-her-to-juvenile-detention

Cole, T. (2017) 'My grandmother's shroud', *The New York Times*, 11 July. Available at: www.nytimes.com/2017/07/11/magazine/my-grandmothers-shroud.html (Accessed: 8 September 2021).

Coupland, N. (2007) *Style: Language Variation and Identity*. Cambridge: Cambridge University Press.

Coupland, N. and Giles, H. (1988) 'The communicative contexts of accommodation', *Anthropology & Education Quarterly*, 8(3–4), pp 175–182.

Dadds, M. (2008) 'Empathetic validity in practitioner research', *Educational Action Research*, 16(2), pp 279–290.

de Sousa Santos, B. and Meneses, M.P. (eds) (2019) *Knowledges Born in the Struggle: Constructing the Epistemologies of the Global South*. New York: Routledge.

Dickinson, E. (1999) *The Poems of Emily Dickinson*. Reading edition. Edited by R.W. Franklin. Cambridge, MA: Belknap Press of Harvard University Press. Available at: www.poetryfoundation.org/poems/56824/tell-all-the-truth-but-tell-it-slant-1263

Douglas, M. (1986) *How Institutions Think*. Syracuse, NY: Syracuse University Press.

Dyson, A.H. and Genishi, C. (eds) (1994) *The Need for Story: Cultural Diversity in Classroom and Community*. Urbana, IL: National Council of Teachers of English. Available at: https://eric.ed.gov/?id=ED365991 (Accessed: 2 September 2021).

Ehret, C. (2018) 'Moments of teaching and learning in a children's hospital: affects, textures, and temporalities', *Anthropology & Education Quarterly*, 49(1), pp 53–71.

Emerson, R.W. (1903) 'Circles', in *The Complete Works of Ralph Waldo Emerson*. Boston, MA: Houghton Mifflin, pp 301–322.

Emerson, R.W. (2003) *Nature and Selected Essays*. Reissue edition. New York: Penguin Classics.

Erickson, M., Hanna, P. and Walker, C. (2020) 'The senior management survey: auditing the toxic university', *Impact of Social Sciences*, 17 February. Available at: https://blogs.lse.ac.uk/impactofsocialsciences/2020/02/17/the-senior-management-survey-auditing-the-toxic-university/ (Accessed: 10 September 2021).

Escott, H. and Pahl, K. (2012) *Language as Talisman: An Annotated Bibliography*. Unpublished manuscript.

Escott, H. and Pahl, K. (2017) 'Learning from ninjas: young people's films as a lens for an expanded view of literacy and language', *Discourse: Studies in the Cultural Politics of Education*, 40(6), pp 803–815.

Escott, H. and Pahl, K. (2019) ' "Being in the bin": affective understandings of prescriptivism and spelling in video narratives coproduced with children in a post-industrial area of the UK', *Linguistics and Education*, 53, pp 1–15.

Facer, K. and Enright, B. (2016) *Creating Living Knowledge: The Connected Communities Programme, Community-University Partnerships and the Participatory Turn in the Production of Knowledge*. Bristol: University of Bristol/Arts and Humanities Research Council. Available at: https://connected-communities.org/index.php/creating-living-knowledge-report/

Facer, K. and Pahl, K. (eds) (2017) *Valuing Interdisciplinary Collaborative Research: Beyond Impact*. Bristol: Policy Press.

Farrell, C.C., Penuel, W.R., Coburn, C., Daniel, J. and Steup, L. (2021) *Research-Practice Partnerships in Education: The State of the Field*. William T. Grant Foundation. Available at: http://wtgrantfoundation.org/library/uploads/2021/07/RPP_State-of-the-Field_2021.pdf

Fine, M. (2016) 'Participatory designs for critical literacies from under the covers', *Literacy Research: Theory, Method, and Practice*, 65(1), pp 47–68.

Fine, M., Bloom, J. and Chajet, L. (2010) 'Betrayal: accountability from the bottom', *Voices in Urban Education*, 30(1), pp 8–19.

Gabel, S. (2005) 'Introduction: disability studies in education', in S. Danforth, S.L. Gabel and S.R. Steinberg (eds) *Disability Studies in Education: Readings in Theory and Method*. New York: Peter Lang, pp 1–20.

Gallacher, L.-A. and Gallagher, M. (2008) 'Methodological immaturity in childhood research? Thinking through "participatory methods"', *Childhood*, 15(4), pp 499–516.

Gilbert, J. (2013) *Common Ground: Democracy and Collectivity in an Age of Individualism*. London: Pluto Press.

Glissant, E. (1997) *The Poetics of Relation*, translated by B. Wing. Ann Arbor, MI: Michigan University Press.

González, N., Moll, L.C. and Amanti, C. (eds) (2005) *Funds of Knowledge: Theorizing Practices in Households, Communities, and Classrooms*. Mahwah, NJ: Lawrence Erlbaum.

Goodman, S. (2003) *Teaching Youth Media: A Critical Guide to Literacy, Video Production, and Social Change*. New York: Teachers College Press.

Goodwin, C. (1994) 'Professional vision', *American Anthropologist*, 96(3), pp 606–633.

Gordon, A.F. (2014) 'On "lived theory": an interview with A. Sivanandan', *Race & Class*, 55(4), pp 1–7.

Gray, P. (2011) 'The decline of play and the rise of psychopathology in children and adolescents', *American Journal of Play*, 3(4), pp 443–463.

Grenfell, M. and Pahl, K. (2019) *Bourdieu, Language-Based Ethnographies and Reflexivity: Putting Theory into Practice*. New York: Routledge.

Guattari, F. (2000) *The Three Ecologies*, translated by I Pindar and P Sutton. London: Athlone Press.

Halberstam, J. (2011) *The Queer Art of Failure*. Illustrated edition. Durham, NC: Duke University Press.

Halberstam, J. (2012) 'Unlearning', *Profession*, pp 9–16.

Hall, R. and Bowles, K. (2016) 'Re-engineering higher education: the subsumption of academic labour and the exploitation of anxiety', *Workplace: A Journal for Academic Labor*, 28, pp 30–47.

Hansen, D.T. (2018) 'Bearing witness to the fusion of person and role in teaching', *The Journal of Aesthetic Education*, 52(4), pp 21–48.

Hansen, D.T. (2021) *Reimagining The Call to Teach: A Witness to Teachers and Teaching*. New York: Teachers College Press.

Hart, A. and Wolff, D. (2006) 'Developing local "communities of practice" through local community–university partnerships', *Planning Practice & Research*, 21(1), pp 121–138.

Heron, J. and Reason, P. (2008) 'Extending epistemology within a co-operative inquiry', in P. Reason and H. Bradbury (eds) *The SAGE Handbook of Action Research*. 2nd edition. London: SAGE, pp 366–380.

Hirshfield, J. (1997) *Nine Gates: Entering the Mind of Poetry*. New York: HarperCollins.

Holland, D., Lachicotte, W., Skinner, D., Cain, C. et al (2001) *Identity and Agency in Cultural Worlds*. Cambridge, MA: Harvard University Press.

Huizinga, J. (1938) *Homo Ludens: Proeve eener bepaling van het spel-element der cultuur* [Homo Ludens: A Study of the Play-Element in Culture]. Haarlem: Tjeenk Willink.

Hull, G.A. and Zacher, J. (2010) 'What is after-school worth? Developing literacy and identity out of school', *Voices in Urban Education*, 30(3), pp 20–28.

Hyatt, D., Escott, H. and Pahl, K. (2017) 'Culture clashes: response to how teachers should respond to non-standard English', in T. Bibby, R. Lupton and C. Raffo (eds) *Responding to Poverty and Disadvantage in Schools: A Reader for Teachers*. London: Palgrave Macmillan, pp 57–76.

Hyland, K. (2009) *Academic Discourse: English in a Global Context*. London: Continuum.

Jamison, L. (2014) *The Empathy Exams*. Minneapolis, MN: Graywolf Press.

Jung, C.G. (1991) *Psyche and Symbol: A Selection from the Writings of C.G. Jung*, translated by R.F.C. Hull, edited by Violet S. de Laszlo. Princeton NJ: Princeton University Press.

Jungnickel, K. (ed) (2020) *Transmissions: Critical Tactics for Making and Communicating Research*. Cambridge, MA: The MIT Press.

Kester, G.H. (2013) *Conversation Pieces: Community and Communication in Modern Art*. Updated edition with a new preface. Berkeley, CA: University of California Press.

Kindon, S., Pain, R. and Kesby, M. (eds) (2010) *Participatory Action Research Approaches and Methods: Connecting People, Participation and Place*. London: Routledge.

Kraftl, P. (2020) *After Childhood: Re-thinking Environment, Materiality and Media in Children's Lives*. New York: Routledge.

Kuby, C.R. and Rowsell, J. (2021) 'Magic(al)ing in a time of COVID-19: becoming literacies and new inquiry practices', *International Studies in Sociology of Education*, pp 1–30. doi:10.1080/09620214.2021.1966826.

Kwan, A. (2011) 'Tycho's talisman: astrological magic in the design of Uraniborg', *Early Science and Medicine*, 16(2), pp 95–119.

Lassiter, L.E. (2005) *The Chicago Guide to Collaborative Ethnography*. Chicago, IL: University of Chicago Press.

Lassiter, L.E. Goodall, H., Campbell, E. and Johnson, M.N. (eds) (2004) *The Other Side of Middletown: Exploring Muncie's African American Community*. Walnut Creek, CA: AltaMira Press

Law, J. (2004) *After Method: Mess in Social Science Research*. London: Routledge.

Leander, K.M., Phillips, N.C. and Taylor, K.H. (2010) 'The changing social spaces of learning: mapping new mobilities', *Review of Research in Education*, 34(1), pp 329–394.

Lemke, J.L. (2000) 'Across the scales of time: artifacts, activities, and meanings in ecosocial systems', *Mind, Culture, and Activity*, 7(4), pp 273–290.

Lesko, N. (2001) *Act Your Age! A Cultural Construction of Adolescence*. New York: Routledge Falmer.

Linde, C. (2009) *Working the Past: Narrative and Institutional Memory*. Oxford: Oxford University Press.

Loveless, N. (2019) *How to Make Art at the End of the World: A Manifesto for Research-Creation*. Durham, NC: Duke University Press.

MacLure, M. (2021) 'Inquiry as divination', *Qualitative Inquiry*, 27(5), pp 502–511.

Manning, E. and Massumi, B. (2014) *Thought in the Act: Passages in the Ecology of Experience*. Minneapolis, MN: University of Minnesota Press.

Massey, D.B. (2005) *For Space*. London: SAGE.

Mauss, M. (2001) *A General Theory of Magic*. 2nd edition. London: Routledge.

McKittrick, K. (2006) *Demonic Grounds: Black Women and the Cartographies of Struggle*. Minneapolis, MN: University of Minnesota Press.

Miles, J. and Springgay, S. (2020) 'The indeterminate influence of Fluxus on contemporary curriculum and pedagogy', *International Journal of Qualitative Studies in Education*, 33(10), pp 1007–1021.

Miller, D. (2009) *Stuff*. Cambridge: Polity.

Mirra, N. (2021) 'Futures bound: re-designing literacy research as a conduit for healing and civic dreaming', *International Studies in Sociology of Education*. https://www.tandfonline.com/doi/full/10.1080/09620214.2021.1945481

Mirra, N., Garcia, A. and Morrell, E. (2015) *Doing Youth Participatory Action Research: Transforming Inquiry with Researchers, Educators, and Students*. New York: Routledge.

Moje, E.B. (2000) '"To be part of the story": the literacy practices of gangsta adolescents', *Teachers College Record*, 102(3), pp 651–690.

Moll, L.C. Amanti, C., Neff, D. and Gonzalez, N. (1992) 'Funds of knowledge for teaching: using a qualitative approach to connect homes and classrooms', *Theory Into Practice*, 31(2), pp 132–141.

Nelson, R. (2013) *Practice as Research in the Arts: Principles, Protocols, Pedagogies, Resistances*. Houndmills: Palgrave Macmillan.

Nygreen, K., Ah Kwon, S. and Sanchez, P. (2006) 'Urban youth building community', *Journal of Community Practice*, 14(1–2), pp 107–123.

Orellana, M.F. (2019) *Mindful Ethnography: Mind, Heart and Activity for Transformative Social Research*. New York: Routledge.

Ottenberg, S. (1990) 'Thirty years of fieldnotes: changing relationships to the text', in R. Sanjek (ed) *Fieldnotes: The Makings of Anthropology*. Ithaca, NY: Cornell University Press, pp 139–160.

Pahl, K. (2002) 'Ephemera, mess and miscellaneous piles: texts and practices in families', *Journal of Early Childhood Literacy*, 2(2), pp 145–166.

Pahl, K. (2014) *Materializing Literacies in Communities: The Uses of Literacy Revisited*. London: Bloomsbury Academic.

Pahl, K. (2016) 'The university as the "imagined other": making sense of community co-produced literacy research', *Collaborative Anthropologies*, 8(1–2), pp 129–148.

Pahl, K. (2019) 'Recognizing young people's civic engagement practices: rethinking literacy ontologies through co-production', *Studies in Social Justice*, 13(1), pp 20–39.

Pahl, K. and Pool, S. (2011) '"Living your life because it's the only life you've got"', *Qualitative Research Journal*, 11(2), pp 17–37.

Pahl, K. and Pool, S. (2017) 'Can we fast forward to the good bits? Working with film: revisiting in the field of practice', in S. Malik, C. Chapain and R. Comunian (eds) *Community Filmmaking: Diversity, Practices and Places*, London Routledge, pp 245–262.

Pahl, K. and Pool, S. (2018) 'Re-imagining artistic subjectivities within community projects', *Open Library of Humanities*, 4(2), pp 1–22.

Pahl, K. and Pool, S. (2021) 'Keeping an eye on the ball: doing research-creation in school', *International Journal of Art and Design Education*, 40(3), pp 655–667

Pahl, K. and Rowsell, J. (2010) *Artifactual Literacies: Every Object Tells a Story*. New York: Teachers College Press.

Pahl, K. Pahl, K., Escott, H., Siebers, J., Steadman-Jones, R., Hurcome, M. and Junior Angling Club, P.A.P. (2017) 'Fishing and youth work, or "what is it about fishing that makes life better"?', in C. Walker, A. Hart and P. Hanna (eds) *Building a New Community Psychology of Mental Health: Spaces, Places, People and Activities*. London: Palgrave Macmillan, pp 83–100.

Paley, V.G. (1986) 'On listening to what the children say', *Harvard Educational Review*, 56(2), pp 122–131.

Patel, L. (2013) *Youth Held at the Border: Immigration, Education, and the Politics of Inclusion*. New York: Teachers College Press.

Patel, L. (2014) 'Countering coloniality in educational research: from ownership to answerability', *Educational Studies*, 50(4), pp 357–377.

Patel, L. (2015) *Decolonizing Educational Research: From Ownership to Answerability*. New York: Routledge.

Patel, L. (2016) *Decolonizing Educational Research: From Ownership to Answerability*. New York: Routledge.

Phillips, R. and Kara, H. (2021) *Creative Writing for Social Research: A Practical Guide*. Bristol: Policy Press.

Pillow, W.S. (2019) 'Epistemic witnessing: theoretical responsibilities, decolonial attitude and lenticular futures', *International Journal of Qualitative Studies in Education*, 32(2), pp 118–135.

Pool, S. (2018) *Everything and Nothing Is Up for Grabs: Using Artistic Methods Within Participatory Research*. University of Bristol (Connected Communities Foundation Series). Available at: https://connected-communities.org/wp-content/uploads/2018/07/Up_For_Grabs_SP.pdf (Accessed: 31 August 2021).

Potter, J. and Cowan, K. (2020) 'Playground as meaning-making space: multimodal making and re-making of meaning in the (virtual) playground', *Global Studies of Childhood*, 10(3), pp 248–263.

Rasool, Z. (2017) 'Collaborative working practices: imagining better research partnerships', *Research for All*, 1(2), pp 310–322.

Ravetz, J. and Ravetz, A. (2017) 'Seeing the wood for the trees: social science 3.0 and the role of visual thinking', *Innovation: The European Journal of Social Science Research*, 30(1), pp 104–120.

Rilke, R.M. (1993) *Letters to a Young Poet*. Revised edition, translated by M.D.H. Norton. New York: W.W. Norton & Company.

Rodriguez Kerr, K., Newhouse, K. and Vasudevan, L. (2020) 'Participatory, multimodal ethnography with youth', in A.I. Ali and T.L. McCarty (eds) *Critical Youth Research in Education*. New York: Routledge, pp 40–59.

Rogers, M. (2012) 'Contextualizing theories and practices of bricolage research', *The Qualitative Report*, 17(48), pp 1–17.

Rowsell, J. and Pahl, K. (2007) 'Sedimented identities in texts: instances of practice', *Reading Research Quarterly*, 42(3), pp 388–404.

Sanjek, R. (ed) (1990) *Fieldnotes: The Makings of Anthropology*. Ithaca, NY: Cornell University Press.

Schneider, A. and Wright, C. (eds) (2006) *Contemporary Art and Anthropology*. London: Berg.

Schneider, A. and Wright, C. (eds) (2010) *Between Art and Anthropology: Contemporary Ethnographic Practice*. Oxford: Routledge.

Scott, S. (2019) *The Social Life of Nothing: Silence, Invisibility and Emptiness in Tales of Lost Experience*. Abingdon: Routledge.

Shannon, D.B. and Truman, S.E. (2020) 'Problematizing sound methods through music research-creation: oblique curiosities', *International Journal of Qualitative Methods*, 19, pp 1–12. https://journals.sagepub.com/doi/10.1177/1609406920903224

Singh, J. (2018) *Unthinking Mastery: Dehumanism and Decolonial Entanglements*. Durham, NC: Duke University Press.

Smith, C. (2021) *How the Word is Passed: A Reckoning with the History of Slavery Across America*. New York: Little, Brown and Company.

Soep, E. and Chavez, V. (2005) 'Youth radio and the pedagogy of collegiality', *Harvard Educational Review*, 75(4), pp 409–434.

Soja, E.W. (2010) *Seeking Spatial Justice*. Minneapolis, MN: University of Minnesota Press.

Sousanis, N. (2015) *Unflattening*. Cambridge, MA: Harvard University Press.

Springgay, S. and Rotas, N. (2015) 'How do you make a classroom operate like a work of art? Deleuzeguattarian methodologies of research-creation', *International Journal of Qualitative Studies in Education*, 28(5), pp 552–572.

Stewart, K. (2007) *Ordinary Affects*. Durham, NC: Duke University Press.

Stewart, K. (2012) 'Precarity's forms', *Cultural Anthropology*, 27(3), pp 518–525.

Street, B.V. (1995) *Social Literacies: Critical Approaches to Literacy in Development, Ethnography and Education*. London: Routledge.

Taylor, F.J. (1979) 'Tench', in R. Walker and L. Moncrieff (eds) *The Shell Book of Angling*. London: David & Charles, pp 158–165.

Thomas, E.E. and Stornaiuolo, A. (2016) 'Restorying the self: bending toward textual justice', *Harvard Educational Review*, 86(3), pp 313–338.

Thomas, K. (1971) *Religion and the Decline of Magic: Popular Beliefs in Sixteenth- and Seventeenth-Century England*. London: Penguin Books.

Thomas-Hughes, H. (2018) 'Ethical "mess" in co-produced research: reflections from a U.K.-based case study', *International Journal of Social Research Methodology*, 21(2), pp 231–242.

Truman, S.E., Hackett, A., Pahl, K., McLean Davies, L. and Escott, H. (2021) 'The capaciousness of no: affective refusals as literacy practices', *Reading Research Quarterly*, 56(2), pp 223–236.

Tuck, E. (2010) 'Breaking up with Deleuze: desire and valuing the irreconcilable', *International Journal of Qualitative Studies in Education*, 23(5), pp 635–650.

Tuck, E. and Yang, K.W. (2014) 'Unbecoming claims: pedagogies of refusal in qualitative research', *Qualitative Inquiry*, 20(6), pp 811–818.

Urquhart, C. (2012) *Grounded Theory for Qualitative Research: A Practical Guide*. Los Angeles, CA: SAGE.

Valenzuela, A. (2017) *Subtractive Schooling: U.S.-Mexican Youth and the Politics of Caring*. Albany, NY: State University of New York Press.

Varenne, H. and McDermott, R. (1998) *Successful Failure: The School America Builds*. Boulder, CO: Westview Press.

Vasudevan, L.M. (2006) 'Looking for angels: knowing adolescents by engaging with their multimodal literacy practices', *Journal of Adolescent & Adult Literacy*, 50(4), pp 252–256.

Vasudevan, L. (2011a) 'An invitation to unknowing', *Teachers College Record*, 113(6), pp 1154–1174.

Vasudevan, L. (2011b) 'Re-imagining pedagogies for multimodal selves', in S. Abrams and J. Rowsell (eds) *Rethinking Identity and Literacy Education in the 21st Century*. Yearbook of the National Society for the Study of Education, 110(1)), pp 88–108.

Vasudevan, L. (2014) 'More than playgrounds: locating the lingering traces of educational anthropology', *Anthropology & Education Quarterly*, 45(3), pp 235–240

Vasudevan, L. (2016) 'Small data, big moments: bearing with-ness in community based research', in paper presented at *Theorizing Our Lives in Critical Research Practices: Exploring Trajectories, Relationships, and Agency Within the Social Contexts of Our Research*, Annual Meeting of the American Educational Research Association, Washington, DC.

Vasudevan, L. and Rodriguez Kerr, K. (2012) 'Re-storying the spaces of education through narrative', in E. Dixon-Román and E.W. Gordon (eds) *Thinking Comprehensively About Education*. New York: Routledge, pp 107–122.

Vasudevan, V., Gross, N., Nagarajan, P. and Clonan-Roy, K. (eds) (2022) *Care-Based Methodologies: Reimagining Qualitative Research with Youth in US Schools*. London: Bloomsbury Academic.

Walton, I. and Cotton, C. (1676/2008) *The Compleat Angler*. Reprint edition, edited by J. Buxton. London: Penguin.

Weber, A. (2017) *Matter and Desire: An Erotic Ecology*. White River Junction, VT: Chelsea Green Publishing.

Wenger, E. (1998) *Communities of Practice: Learning, Meaning, and Identity.* Cambridge: Cambridge University Press.

Willats, S. (2012) *Artwork as Social Model: A Manual of Questions and Propositions.* Sheffield: Research Group for Artists Publications.

Winn, M.T. (2015) 'Exploring the literate trajectories of youth across time and space', *Mind, Culture, and Activity,* 22(1), pp 58–67.

Woolf, V. (1978/1941) *Between the Acts.* London: Granada Publishing.

Wortham, S. and Rhodes, C. (2012) 'The production of relevant scales: social identification of migrants during rapid demographic change in one American town', *Applied Linguistics Review,* 3(1), pp 75–99.

Wortham, S., Nichols, B., Clonan-Roy, K. and Rhodes, C. (2020) *Migration Narratives: Diverging Stories in Schools, Churches, and Civic Institutions.* London: Bloomsbury Academic.

Yoon, H.S. and Templeton, T.N. (2019) 'The practice of listening to children: the challenges of hearing children out in an adult-regulated world', *Harvard Educational Review,* 89(1), pp 55–84.

Younge, G. (2016) *Another Day in the Death of America: A Chronicle of Ten Short Lives.* New York: Bold Type Books.

Index

References to footnotes show both the page number and the note number (231n3).